Getting Started with PICs
Volume 2

A Collection of 2007 Nuts & Volts Magazine Articles

The publisher offers special discounts on bulk orders of this book.

For information contact:

Electronic Products
P.O. Box 251
Milford, MI 48381
www.elproducts.com
chuck@elproducts.com

The Microchip name and logo, MPLAB® and PIC® are registered trademarks of Microchip Technology Inc. in the U.S.A. and other countries. PICkit™ is a trademark of Microchip Technology Inc. in the U.S.A. and other countries.

BasicAtom is a trademark of Basic Micro Inc.
Basic Stamp is a trademark of Parallax Inc.

All other trademarks mentioned herein are property of their respective companies.

Printed in United States of America
Cover design by Rich Scherlitz

Table of Contents

Introduction

In late 2005 Nuts & Volts Magazine offered me the chance to write a column about getting started with Microchip PIC® MCU's. I had been writing freelance articles and even published a couple books prior to this offer but it was still very exciting to get an offer like this. I released a book in 2008 titled "Getting Started with PICs" which is a collection of the first 12 articles published in January 2006 through December 2006. This book you are reading is the second set of 12 articles published in January 2007 through December 2007.

The articles in this book jump around a bit from using the PICBASIC PRO compiler, to using the Basic Atom Microcontrollers to introducing the C language. There is a lot of useful information in this collection of articles. They are mostly reprints of the Nuts & Volts magazine articles but some minor changes and improvements are included that might have missed the magazines publishing deadline.

If you are a regular reader of the column or joined it late, then this book will make a nice reference for columns you already read or columns you might have missed. I myself reference my own articles often so having them in a bound book form is very handy.

The column continues on into 2009. In 2008 I finally outlined my direction and focused on the true beginner. This has been a great success based on reader feedback. In 2008 the column also changed to an every other month column which reduced the number of articles to six per year. I'll be combining the 2008 and 2009 articles into a book and releasing that in the future to keep this series going.

I hope you gain from the information in this book. Please send me your feedback to my email at chuck@elproducts.com.

Chuck Hellebuyck

January 2007 - PIC12F675 Replaces the 555 Timer

If you have been fooling with electronics as long as I have you will have seen a lot of changes throughout the years. I watched the birth of integrated circuits which led to Op-Amps and TTL or CMOS digital chips and eventually the Microcontroller. Somewhere along the way an 8-pin chip used by hobbyists all over the world was developed. It was known as the 555 timer (shown as a block diagram in Figure 1). This little 8-pin chip would be the main building block of many electronic projects. Even Forrest Mims who wrote most of those Radio Shack project books many years ago used the 555 in various applications. By just adding a few external capacitors and resistors plus an occasional diode, you could make the 555 chip do amazing things. Though some still use the 555 I've found that the 8-pin PICmicro's can do everything a 555 can do and more.

Many years ago Microchip released the first 8-pin microcontroller and I was blown away. To have all the power of a PIC inside a small eight pin package was just amazing. I began to think of all the 555 applications it could replace and do so much more yet not take up any more space. In fact it would take less space since all the external capacitors, resistors and diodes were replaced by software. The early eight pin PICs were limited in their features but still far more powerful than a 555. As the years have gone by, more powerful eight pin PICs have been released. The only problem with all this was you needed to know how to program it in order to use them. That was a huge hurdle for many people including me. But with the release of PicBasic Pro and other PIC compilers, low cost PIC programmers such as my EZPIC programmer and lots of sample code for PICs, it's not nearly as difficult to use a PIC in place of a 555.

555 Timer Block Diagram

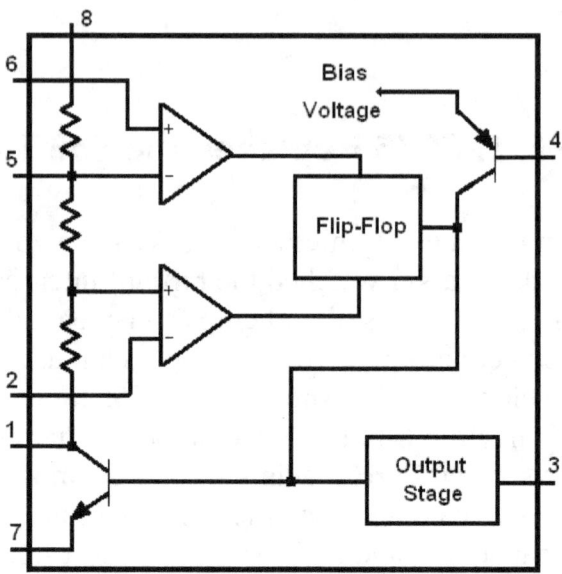

Figure 1

One of the 555 projects I built was a large LED countdown clock for my nephew about 15 years ago that had a 555 oscillator at the heart of it. He wanted this for his annual roller hockey birthday party. He was 11. I knew it wasn't very accurate since it was really based on a resistor-capacitor charging scheme that is greatly affected by temperature and voltage variation but it was good enough for a street hockey game. I built it with a potentiometer so I could adjust the accuracy but never showed him how to do that. Then one day, several months after the party, he used it time a speech he had to do for school. He called me in panic because he thought he had broken something. The clock was really inaccurate. I had to explain to him about how it worked and the reason for its inaccuracy and comforted him that he didn't break it. I was more surprised that he used it and thought to myself how I should have built it with a crystal based oscillator instead. It wasn't long after that that I started getting into PICs.

He recently told me he still has that old thing which got me thinking. With an eight pin PIC and crystal accuracy, I could build a really accurate time base. I could even add temperature compensation using an A/D port to read a temperature sensor. It also got me thinking about all those old 555 circuits I use to build such as a one-shot (monostable multivibrator) and clock oscillator (astable multivibrator). I even found an old Forrest Mims circuit book that used the 555 to drive LED's in sequence through a 7441 decoder driver TTL IC. All these are so easy to do with a

PIC I realized that there really wasn't any need to use a 555 and it had, in my mind, been replaced by this eight pin PIC.

Figure 2

For this article I thought I would show how to build a simple 555 one-shot replacement circuit using an external interrupt. I will also have it act as a pulse generator sending a 50% duty cycle signal out another output pin at the same time to cover another 555 goodie. These were popular 555 projects in their day. For both of these I will use the PicBasic Pro compiler and the PIC12F675 8-pin PIC (Figure 2).

You can get a PIC12F675 for around $1.00 in small quantity at various sources including Microchipdirect.com. You can even get a free sample from Microchip. I like this part since it's a 14-bit core PIC which is the same as the 16F876A that I like to use in my Ultimate OEM module and has the extra stack levels. Most compilers including PicBasic Pro offer more commands on PICs with larger stacks. Some of the other 8-pin PICs Microchip offers are 12-bit core PICs and only have a smaller stack. PicBasic Pro will work with these but some commands are not available.

Another big advantage to the PIC12F675 is the on-board Analog to Digital (A/D) convertor. The 12F675 also has a comparator, 128 bytes of internal EEPROM, 64 bytes of RAM for variables and 1k of Flash memory for the program. It even has a built-in 4 mhz oscillator option so you don't need an external resonator to run it. On all the 8-pin PICs, the MCLR pin has the option to internally pull the reset line high allowing that pin to be used as an I/O pin also.Therefore this is a PIC that only requires 5 volts and ground to make it run.

All these features add complexity to the setup in the software though so that will take a little explanation, especially the internal oscillator. So lets get that out of the way.

Internal Oscillator

The 12F675 internal oscillator is built right into the silicon wafer that makes up the chip. The factory then calibrates the oscillator during the assembly process and they store a correction value at the last location of program memory. You need to use that value to get the most accurate clock possible. That requires you to be a little careful when programming it for the first time. I used my EZPIC programmer, that I wrote about last year at this time, to program the PIC12F675. The software used is the ICPROG.exe freeware from IC-Prog.com. This software will first read the PIC12F675 to see what the value of the last location in memory is set to. You can see in Figure 3 the value for the PIC I loaded.

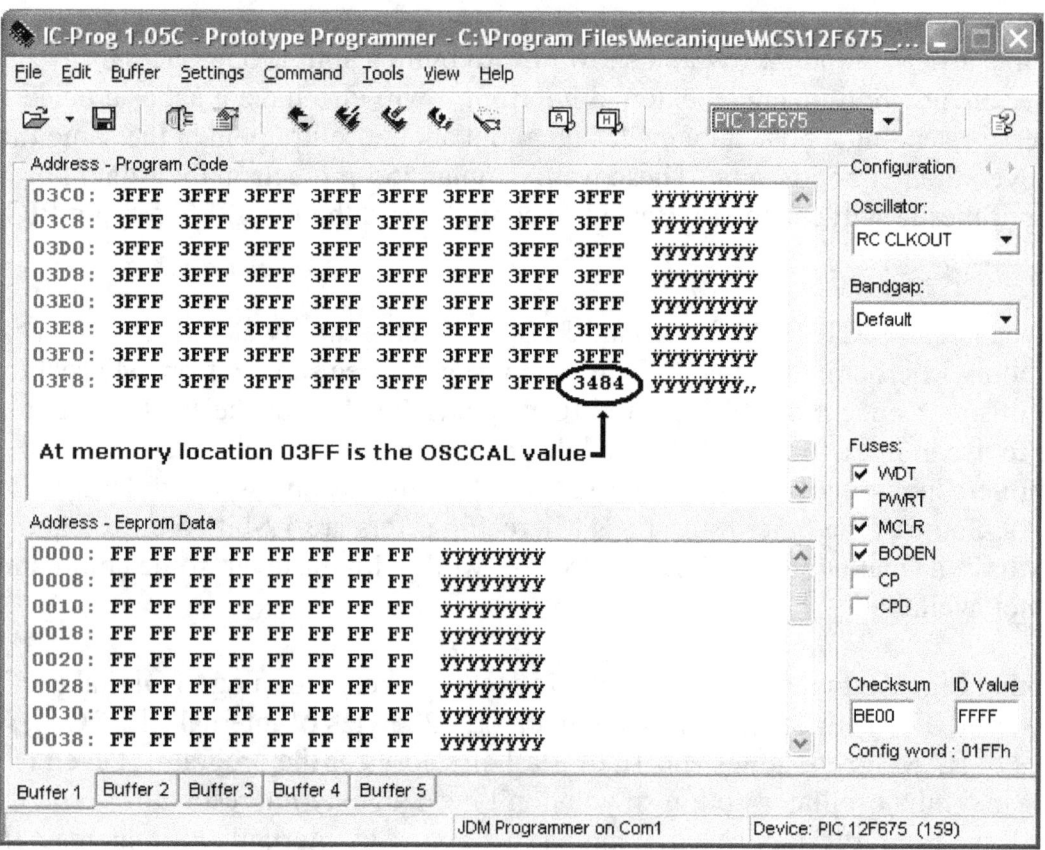

Figure 3

If you write over that location, the calibration adjustment value will be lost. I recommend reading the chip and then writing it down on a small sticker and stick it to the bottom of the chip. This way you have the value even if it's erased. The ICPROG.exe software will even warn you before programming as seen in figure 4.

10

After it reads the PIC memory it will ask if you want to use the calibration value read or the value in the .hex file you are trying to program into the PIC12F675. You can choose which one you want to use. In my program I didn't set that location to anything so it showed up as 3FFF. Therefore I chose to use the value read not the 3FFF value. The value is 3489 on this PIC12F675 while the previous screen in figure 3 shows 3484. This is because I used two different PICs when I created these screen shots. This also shows how close the calibration values are from PIC to PIC. If you ever accidentally erase one you can just read a different PIC12F675 and at least be close.

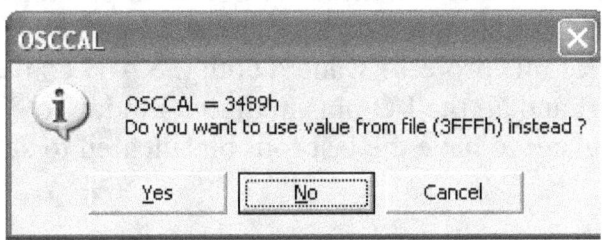

Figure 4

You need to store that value in the OSCCAL register within the PIC to make the internal oscillator operate properly. PicBasic Pro handles this for you and I'll show you that in the code later. You also have to configure the PIC for internal oscillator when it's programmed. You also have to configure the MCLR pin for internal or external operation. You do that in the configuration register also at programming time. PicBasic Pro has an include file that automatically establishes this. It's called 12F675.inc and it's in the PicBasic Pro directory. I modified it to match the format below.

```
NOLIST
ifdef PM_USED
    LIST
    include 'M12F675.INC'        ; PM header
    device  pic12F675, intrc_osc, wdt_on, mclr_off, protect_off
    XALL
    NOLIST
else
    LIST
    LIST p = 12F675, r = dec, w = -302
    INCLUDE "P12F675.INC"     ; MPASM  Header
    __config _INTRC_OSC_NOCLKOUT & _WDT_ON & _MCLRE_OFF & _CP_OFF
    NOLIST
endif
    LIST N
```

The line INTRC_OSC_NOCLKOUT sets the internal oscillator mode. The MCLRE_OFF sets the MLR pin to internal operation. There are a few other special registers I wanted to explain before jumping into the code.

Special Register Setup
Because the PIC12F675 has few pins and lots of features, the various functions such as comparator and A/D are multiplex connected to the actual metal pin. To select which ones get connected requires you to set them up in software. The ANSEL (analog select) register and the CMCON (comparator control) register need to be setup at the beginning of your program so you know how the PIC is initially connected. For this project I wanted both the A/D converter and comparator turned off and all the I/O pins in digital mode. To do that I first have to setup the ANSEL register to have the last four bits cleared to zero as referenced in Figure 5.

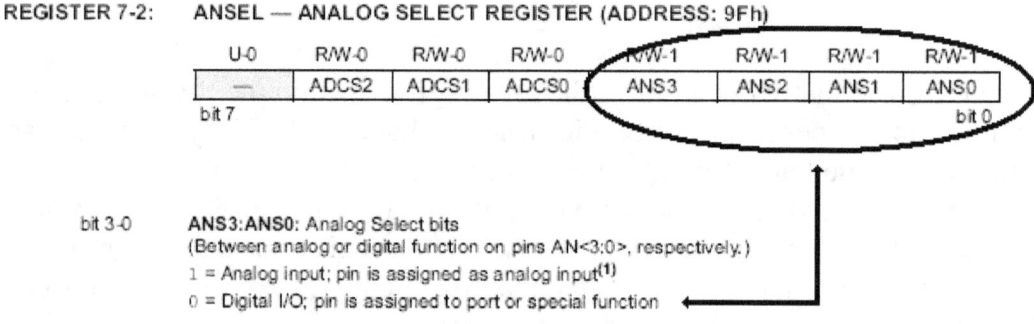

Figure 5

The comparator requires the CMCON register to be setup for digital operation by setting the last three bits to their proper setting. In this case setting them all to one disconnects the comparator and lets the pins be digital I/O and seen in Figure 6.

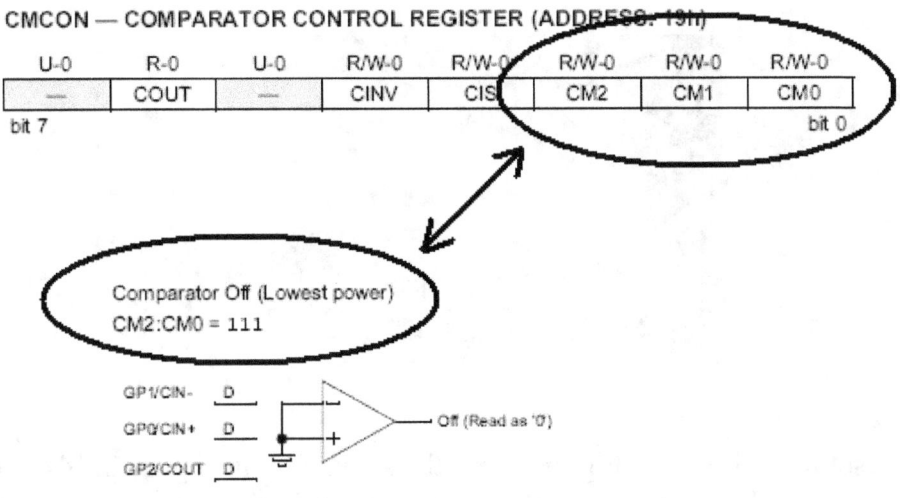

Figure 6

This probably seems a little confusing at first and this is all in the PIC12F675 data sheet along with a lot more information. This is all you really need to know though to get started. The first project is always the hardest.

How it Works

The project will just flash the green LED on GP0 at close to 5 Hz while waiting for the button to be pressed. This is a 5 Hz 50% duty cycle clock much more accurate than the one I built for my nephew. When the button is pressed a one-shot 250 msec pulse will be output on the GP1 pin thus briefly lighting the red LED and freezing the green LED in whatever state it is during that 250 msec. This will throw off the clock accuracy but I really just wanted to show how easy it was to do these functions in software. The one shot output could be used to reset something so the clock may not need to be accurate at that point anyway.

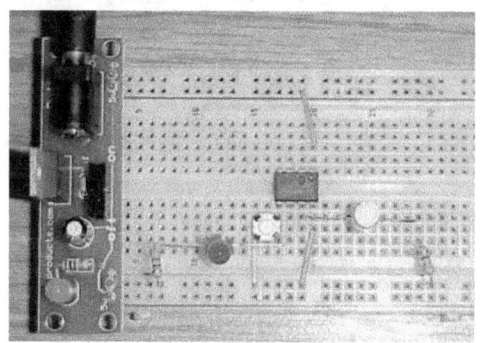

Figure 7

Hardware

The whole setup is shown in Figure 7 and the schematic in Figure 8. It's really a simple setup. I used one of my 5-volt regulator breadboard modules to supply power to the rails. You could replace this with the 7805 circuit shown in the schematic. The project uses the internal pullups of the PIC to keep the switch input on the interrupt pin GP2 high. The interrupt is set to trip on a falling edge. These are both setup in the software OPTION register. The PIC directly drives the LEDs.

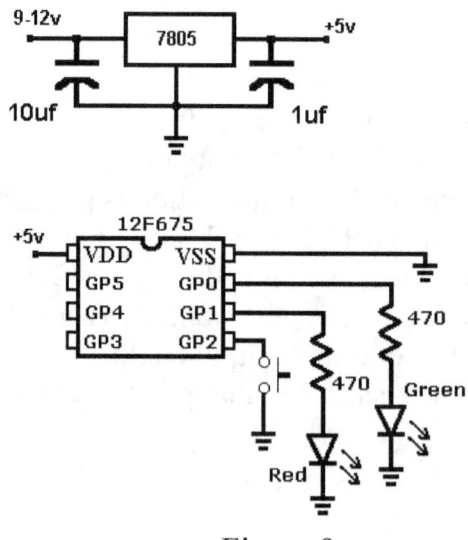

Figure 8

Software Listing

```
'**************************************************
'* Name    : 12F675_555test.BAS                  *
'* Author  : Chuck Hellebuyck                     *
'* Notice  : Copyright (c) 2006 Electronic Products *
'*         : All Rights Reserved                  *
'* Date    : 11/1/06                              *
'* Version : 1.0                                  *
'* Notes   :                                      *
'*         :                                      *
'**************************************************


DEFINE OSC 4
DEFINE OSCCAL_1K 1

CMCON = 7
ANSEL = 0
TRISIO = %00111100          'GP2-5 Inputs, GP1,0 Outputs
WPU = %00000100             'GP2 Pullup Enabled
GPIO = 0                    'All Ports Low to Start
OPTION_REG = %00000000      ' Internal Pull-ups enabled, Trigger on
                            ' Falling Edge
INTCON = %10010000          'Enable External Interrupt


ON INTERRUPT GOTO pulse  ' Create Interrupt

Main
   High 0          'Send continuous 5 Hz
   pause 100       ' 50% duty cycle
   low 0           ' clock pulse out
   pause 100       ' pin GP0
goto Main

' *** Interrupt handler routine ***
DISABLE
pulse:
   high 1          ' Send 250 msec pulse
   pause 250       ' out the GP1 pin
   Low 1
```

```
INTCON.1 = 0       ' Clear interrupt flag
RESUME
ENABLE
```

Software (How It Works)

The software starts out very similar to other code I've written. I first define the oscillator speed at 4 Mhz. Then issue the special command that takes the OSCCAL value at the end of program memory and stores it into the OSCCAL register. This is a special PicBasic Pro DEFINE that is explained in the early part of the PicBasic Pro manual.

```
DEFINE OSC 4
DEFINE OSCCAL_1K 1
```

Next those special register setups are established as we talked about earlier. Comparators are turned off and the A/D is not connected to the outside pins.

```
CMCON = 7
ANSEL = 0
```

Now the I/O is setup. The 8-pin PICs use a different name for the I/O so the register to control is the TRISIO register. Making a bit a 1 makes it an input and a 0 makes it an output. Since there are only six I/O, the left most bits don't matter what they are set to.

```
TRISIO = %00111100          'GP2-5 Inputs, GP1,0 Outputs
```

There is a special register just for the pull-up resistors. It's the WPU (weak pullup) register. A one enables the pull-up and a 0 disables it. This is nicer than the 16F876A since you can set each pull-up individually here but the 16F876A is all on or all off. This doesn't turn them on yet. That is done in the OPTION register.

```
WPU = %00000100             'GP2 Pull-up Enabled
```

I initialize all the I/O to low by directly writing to the GPIO port. Now looking at this I should have done this before setting the TRISIO register. It will work but occasionally the LEDs may initially flash on.

```
GPIO = 0                    'All Ports Low to Start
```

The OPTION register does a lot of things but here we only need to turn on the internal pull-ups and set the direction for the interrupt to falling edge (high to low).

```
OPTION_REG = %00000000   ' Internal Pull-ups enabled, Trigger on
                         ' Falling Edge
```

The interrupt control register enables only the external interrupt by setting the 5th bit. The 8th bit turns all interrupts on.

```
INTCON = %10010000       'Enable External Interrupt
```

The ON INTERRUPT command is issued to tell the program where to go when an external interrupt occurs. This command makes interrupts very easy.

```
ON INTERRUPT GOTO pulse      ' Create Interrupt
```

The main loop is simple. Just a simple flash an LED loop we've done before.

```
Main
    High 0          'Send continuous 5 Hz
    pause 100       ' 50% duty cycle
    low 0           ' clock pulse out
    pause 100       ' pin GP0
goto Main
```

The interrupt routine is also quite easy. We first DISABLE any future interrupts while the handler is functioning. This is hardly an issue since it will run a lot faster than a person can press the switch but this also helps to not react to switch bounce multiple times.

```
' *** Interrupt handler routine ***
DISABLE
```

The interrupt handler just pulses the GP1 pin for 250 milliseconds. This is where you can easily adjust the one-shot timing without having to change resistors or capacitors like you would with a 555.

```
pulse:
```

```
high 1          ' Send 250 msec pulse
pause 250       ' out the GP1 pin
Low 1
```

We have to clear the interrupt flag before leaving the interrupt and then issue the RESUME and ENABLE commands.

```
INTCON.1 = 0    ' Clear interrupt flag
RESUME
ENABLE
```

Short and simple and our 555 one-shot replacement with a side of 5 Hz oscillator is ready to go. This took so little code space. Only 98 of 1024 words or program memory were used.

Conclusion

As you can see the PIC12F675 is a handy little part. PicBasic Pro comes through again to make the software easy. This program can be expanded in so many ways. I just wanted to give the reader a base to build more complex 12F675 projects from. With the SEROUT/SERIN command in PicBasic Pro you can make any pin a serial communication pin. Add a large memory 8-pin EEPROM using the I2COUT/I2CIN commands to a 12F675 and you have memory storage. Add a temperature sensor to an A/D pin and you have only used 5 of the 6 pins.

Now one thing I did not mention is the I/O limitation. If you look in the data books and you will see that the GP3 pin is an input only pin so you really have only 5 outputs and 6 inputs with this chip. Five is all you need though to make a small temperature data logger that I just described. Give it a try and let me know through email how it worked. Send any comments about the articles to chuck@elproducts.com. I like getting the feedback. You can also visit www.beginnerelectronics.com for the EZPIC programmer and the 5v regulator breadboard module. I have a whole bunch of new ideas for 2007 so stay tuned.

February 2007 - Quick and Dirty Hockey Scoreboard

Once you learn how to control an LCD module suddenly you see them all over the place and numerous ideas for LCD projects come to mind. Over the years I've done many projects with LCDs and the whole LCD business has grown to the point that just about every kind of LCD you could imagine is available. Serial LCDs that receive information in RS232 style format are popular and have come down in price. Parallel LCDs which require a little more software to drive are popular because they can be bought in surplus for around $10. LED backlight LCDs are also readily available in lots of different colors. I've also received a lot of email from new readers to the column asking for more LCD examples. I pulled a little project out of my library for this month to show how having a development board with built in LCD can really come in handy. I also show that the Atom software makes driving one of the lower cost parallel LCDs is as easy as driving a serial LCD because of the LCDWRITE command.

Scoreboard
Several years ago I developed one of my most favorite Atom development boards that I call the BasicBoard. I tried to put all the features in one board that I use most often in my projects. The BasicBoard has switches, LEDs, speaker, potentiometer, serial port, expansion ports and of course an LCD. I used a 40 pin Atom interpreter chip at the center of it which is a PIC16F877A with the Atom firmware installed. This board came in handy when I needed to build a quick timer for a mini-stick hockey game I promised to play with my son.

Figure 1

If you're not familiar with mini-stick hockey it's when kids play hockey on their knees using those little plastic souvenir hockey sticks and a soft round ball. He often just plays until someone gets to 10 or 20 goals but my knees can't take that so I wanted to put a time limit on it. I didn't have time to build a whole setup so I pulled out a BasicBoard and in a short time had a whole scoreboard developed as seen in Figure 1. I wasn't real worried about accuracy of the time but since the BasicBoard used a 20 Mhz resonator and I was only looking for a 20 minute timer, this setup was accurate enough. In fact, I didn't even need to use a timer interrupt, I could just use a simple PAUSE command to get the right timing.

I got a little carried away and added score to the setup since it didn't take much more effort and I had the space on the LCD. Having the switches and LEDs also pre-connected made this simply a software project. The original even had a second timer to tell when to change lines in case he had a group of kids playing that were taking turns. I removed that for this article so this would fit in a reasonable amount of magazine space.

Software
```
'******************************************************************
'*  Name     : scoreboard2.BAS                              *
'*  Author   : Chuck Hellebuyck                                *
'*  Notice   : Copyright (c) 2004 Electronic Products         *
'*            : All Rights Reserved                           *
'*  Date     : 7/6/2004                                       *
'*  Version  : 1.2                                            *
```

```
'*   Notes     :        This program is demonstrates how to use           *
'*            :        the 40 pin Atom Interpreter                        *
'*            :        chip and the BasicBoard to build a scoreboard.     *
'***************************************************************


spkr con 20              'Initialize Speaker Connection
tns var byte             '0x:xx 10's digit display variable
mns var byte             'x0:xx 1's digit display variable
scs var byte             'xx:0x 1/10's digit display variable
ths var byte             'xx:x0 1/100's digit display variable
mn var byte              'total minutes varible for clock
sc var byte              'total seconds varible for clock
hscore var byte          'Home Score variable
vscore var byte          'Visitor Score variable

' *** Initialize LCD to 2x16 *********
pause 500
lcdwrite 17\16,outc,[initlcd1,initlcd2,twoline,scrblk,clear,home]

' *** Initialize Section ***********
main

' *** Display Banner of "ScoreBoard" for 1 second ****
lcdwrite 17\16,outc,[clear,home,"ScoreBoard"]
pause 1000

' *** Initialize clock to 20:00 and Home 0 vs Visitor 0 ***
init
mn = 20
sc = 0
hscore = 0
vscore = 0

' *** Create timer values to be displayed ***
tns = mn/10          ' Ten's digit is minutes divided by 10
mns = mn//10         ' One's digit is the remainder
scs = sc/10          ' Tenths digit is seconds divided by 10
ths = sc//10         ' Hundreths is the remainder
```

' *** Display the time on the LCD ***
lcdwrite 17\16,outc,[Clear,home,"HOME ",dec tns,dec mns,":",dec scs,dec ths," VIS"]
lcdwrite 17\16,outc,[scrram+$40+1,dec hscore,scrram+$40+14,dec vscore]

' *** Main Loop of Code ***
loop

```
pause 975                    ' Adjustment for accuracy of clock
if sc <> 0 then              ' Test if seconds is not zero
sc =sc - 1                   ' Reduce seconds by one
elseif sc = 0 and mn > 0     ' Seconds is zero so test minutes also
sc = 59                      ' Reset secondsto 59
mn = mn - 1                  ' Reduce minutes by one
endif


tns = mn/10                  ' Ten's digit is minutes divided by 10
mns = mn//10                 ' One's digit is remainder
scs = sc/10                  ' Tenth's digit is seconds divided by 10
ths = sc//10                 ' Hundreth's digit is remainder
```

' *** Update Display **********
lcdwrite 17\16,outc,[Clear,home,"HOME ",dec tns,dec mns,":",dec scs,dec ths," VIS"]
lcdwrite 17\16,outc,[scrram+$40+1,dec hscore,scrram+$40+14,dec vscore]

'** Check for home score ***
```
if in19 = 0 then             'If P19 switch is pressed
hscore = hscore + 1          ' then increase home score
```

'** Update Display ***********
lcdwrite 17\16,outc,[Clear,home,"HOME ",dec tns,dec mns,":",dec scs,dec ths," VIS"]
lcdwrite 17\16,outc,[scrram+$40+1,dec hscore,scrram+$40+14,dec vscore]
endif

'*** Wait for switch to be released ***
holdh

```
if in19 = 0 then holdh

'*** Check for VIS score ***
if in12 = 0 then                    'If P12 switch is pressed
vscore = vscore + 1                 ' then increase visitor score

'*** update display ************
lcdwrite 17\16,outc,[Clear,home,"HOME  ",dec tns,dec mns,":",dec scs,dec ths,"
VIS"]
lcdwrite 17\16,outc,[scrram+$40+1,dec hscore,scrram+$40+14,dec vscore]
endif

'*** Wait for switch to be released ***
holdv
if in12 = 0 then holdv


'*** Freeze Time ***

if in18 = 0 then                    'If Switch 18 is pressed then
high 7                              ' light LED7 and
holdt                               ' hold up the clock
if in13 = 1 then holdt              ' until switch is pressed
low 7                               'Switch pressed LED7 off
endif

' *** Check for Main Clock Time up ***
if sc = 0 and mn = 0 then           ' Test for time equal to zero
high 7                              ' time up light LED7
sound spkr, [8000\200]              ' Play time up tone for 8 seconds
tuhold                              'Hold up program until switch 13
if in13 = 1 then tuhold             ' is pressed as reset switch
low 7                               ' LED 7 off
mn = 20                             ' Reset time to 20:00
sc = 0

goto loop                           'Loop back to do it all again
endif

goto loop                           'Final loop if all else passes
```

How It Works

Using the CON (constant) and VAR (variable) directive all the variables are established. The first being the speaker connection. It's easier to understand "spkr" is speaker than p20.

```
spkr con 20                'Initialize Speaker Connection
```

Then the variables required to store the time and the display variables are reserved in RAM using the VAR directive. This program only needs byte size variables.

```
tns var byte               '0x:xx 10's digit display variable
mns var byte               'x0:xx 1's digit display variable
scs var byte               'xx:0x 1/10's digit display variable
ths var byte               'xx:x0 1/100's digit display variable
mn var byte                'total minutes varible for clock
sc var byte                'total seconds varible for clock
hscore var byte            'Home Score variable
vscore var byte            'Visitor Score variable
```

The LCD needs to be initialized as a 2x16 LCD with the cursor block visible and blinking (scrblk). This command also clears the display (clear) and puts the cursor at the first position (home). We also pause 500 milliseconds before this command to allow the LCD hardware to power up properly.

```
' *** Initialize LCD to 2x16 *********
pause 500
lcdwrite 17\16,outc,[initlcd1,initlcd2,twoline,scrblk,clear,home]
```

Next the program shows the display is working by displaying "ScoreBoard" on the first line and delay for 1 second so we will see it.

```
' *** Initialize Section ***********
main

' *** Display Banner of "ScoreBoard" for 1 second ****
lcdwrite 17\16,outc,[clear,home,"ScoreBoard"]
pause 1000
```

The program also has to initialize the clock and score and this is done by initializing the variables.

```
' *** Initialize clock to 20:00 and Home 0 vs Visitor 0 ***
init
mn = 20
sc = 0
hscore = 0
vscore = 0
```

The time is actually broken out from the two time variables by math functions. We get the first digit by dividing by 10 then use the remainder for the ones digit. We do the same for the two digits beyond the decimal point. This is a simple way to break up a number into individual digits to be displayed on an LCD.

```
' *** Create timer values to be displayed ***
tns = mn/10          ' Ten's digit is minutes divided by 10
mns = mn//10         ' One's digit is the remainder
scs = sc/10          ' Tenths digit is seconds divided by 10
ths = sc//10         ' Hundreths is the remainder
```

The calculated data is displayed on the LCD. The second line is controlled by the SCRRAM+$40 command. The second line position starts at 40 hex on the LCD character map inside the LCD display as seen in Figure 2.

00	01	02	03	04	05	06	07	08	09	10	11	12	13	14	15	16	17	18	19	← Character position (dec.)
00	01	02	03	04	05	06	07	08	09	0A	0B	0C	0D	0E	0F	10	11	12	13	← Row0 DDRAM address (hex)
40	41	42	43	44	45	46	47	48	49	4A	4B	4C	4D	4E	4F	50	51	52	53	← Row1 DDRAM address (hex)
14	15	16	17	18	19	1A	1B	1C	1D	1E	1F	20	21	22	23	24	25	26	27	← Row2 DDRAM address (hex)
54	55	56	57	58	59	5A	5B	5C	5D	5E	5F	60	61	62	63	64	65	66	67	← Row3 DDRAM address (hex)

Figure 2

The position you want is then just added to the value. Notice I add "14" to position the visitor score and it is treated as decimal 14 not a hex value. This is because the Atom compiler doesn't see a '$' in front of it so it converts it and does the math. Nice little feature.

```
' *** Display the time on the LCD ***
lcdwrite 17\16,outc,[Clear,home,"HOME  ",dec tns,dec mns,":",dec scs,dec ths," VIS"]
lcdwrite 17\16,outc,[scrram+$40+1,dec hscore,scrram+$40+14,dec vscore]
```

The main loop is entered next and this just decrements the clock and updates the display in a similar fashion to what we did earlier. The only difference is the clock decrement calculation. Also notice the "pause 975" at the top of the loop. This is a crude adjustment for accuracy to make the loop close to 1 second. I just timed it next to a stopwatch via trial and error to get that number. In fact its probably a little off since I removed a little code from the original program for this article. You will have to adjust that a little by making it slightly larger.

```
' *** Main Loop of Code ***
loop

pause 975                     ' Adjustment for accuracy of clock
if sc <> 0 then               ' Test if seconds is not zero
sc =sc - 1                    ' Reduce seconds by one
elseif sc = 0 and mn > 0      ' Seconds is zero so test minutes also
sc = 59                       ' Reset secondsto 59
mn = mn - 1                   ' Reduce minutes by one
endif
```

The "Home" score and "Visitor" score are next and all these sections do is look for a switch to be pressed. If one of the switches is pressed, the score is incremented and then waits for you to release the switch. Now this is actually lousy code writing because holding the button actually stops the clock. It also requires you to hold the switch for almost a second because of that long Pause 975 delay above. This is the cost of doing something quick and dirty so my son wasn't waiting forever for me to finish the scoreboard. If I would have used a timer interrupt as the clock base I could have made the main loop shorter and scanned the keys quicker without ever messing up the clock accuracy.

```
'** Check for home score ***
if in19 = 0 then                    'If P19 switch is pressed
hscore = hscore + 1         ' then increase home score

'** Update Display  ***********
lcdwrite 17\16,outc,[Clear,home,"HOME  ",dec tns,dec mns,":",dec scs,dec ths," VIS"]
lcdwrite 17\16,outc,[scrram+$40+1,dec hscore,scrram+$40+14,dec vscore]
endif

'*** Wait for switch to be released ***
holdh
if in19 = 0 then holdh

'*** Check for VIS score ***
if in12 = 0 then                    'If P12 switch is pressed
vscore = vscore + 1         ' then increase visitor score

'*** update display  *************
lcdwrite 17\16,outc,[Clear,home,"HOME  ",dec tns,dec mns,":",dec scs,dec ths," VIS"]
```

```
lcdwrite 17\16,outc,[scrram+$40+1,dec hscore,scrram+$40+14,dec vscore]
endif
```

```
'*** Wait for switch to be released ***
holdv
if in12 = 0 then holdv
```

The next section uses that stop the clock fault of the score setting to actually stop
the clock. It reads the switch the same way it read the score switches but this time
holds until the switch is pressed again. Switch bounce could be a problem here but
the time to set the LED allows this to work well. If this wasn't the case, it could
actually see two presses quickly because the contacts of momentary switches can
internally bounce causing multiple contacts quickly which can be read as two or
more presses. The Atom is a little slower than a compiled PIC so this is less of an
issue.

```
'*** Freeze Time ***

if in18 = 0 then              'If Switch 18 is pressed then
high 7                        ' light LED7 and
holdt                         ' hold up the clock
if in13 = 1 then holdt        ' until switch is pressed again
low 7                         'Switch pressed LED7 off
endif
```

This last section tests the clock values to see if they are at zero. This indicates time
up so the SOUND command is used to drive a tone from the speaker. The program
also pauses the program after the tone is complete and waits for the user to press
the P13 switch. When the switch is pressed the program jumps back to the top and
resets the clock to 20:00 and clears the score. The timer starts running again. If the
clock isn't at zero then this section is skipped and the program jumps to the loop
label.

```
' *** Check for Main Clock Time up ***
if sc = 0 and mn = 0 then          ' Test for time equal to zero
high 7                             ' time up light LED7
sound spkr, [8000\200]             ' Play time up tone for 8 seconds
tuhold                             'Hold up program until switch 13
if in13 = 1 then tuhold            ' is pressed as reset switch
low 7                              ' LED 7 off
mn = 20                            ' Reset time to 20:00
sc = 0
goto loop                          'Loop back to do it all again
```

```
endif

goto loop                          'Final loop if all else passes
```

There are lots of areas to shorten this code. All the display commands could be combined into a subroutine and so could some of the calculations of display digits but shorter code wasn't my goal, a quick solution to save my knees was all I was looking for.

Hardware

The schematic for the BasicBoard is shown in Figure 3. The schematic shows four momentary switches pre-wired to the Atom PIC with external pull-up resistors. I put eight LEDs all connected to PortB (P0 thru P7) through a 330 ohm resistor bank. The speaker is prewired to P20 pin through a 10 uf cap which is an easy way to create sound as shown in the December 2006 article. The LCD is a simple 4-bit connection scheme. I don't have it wired for backlight operation.

This setup is easily built on a breadboard if you have the components lying around. I personally like laying out these types of development boards. Call me crazy but I find it a lot of fun. It costs more initially but saves time in the long run. I do get many emails asking me which development board I recommend and frankly I don't really know. I like mine since I designed them but everybody has their own preference. In most cases you are spending a lot of money for these boards so you want as many options as possible included I guess but even that becomes a problem because you may have too many options that require all sorts of jumpers and more difficult software setup. You also end up paying for features you may never use.

My Ultimate OEM modules resulted from all that experimenting which is just a very simple development module. My point is don't be afraid to shop around to find what you think will work for you and then have a couple around for those simple projects like the Scoreboard. Just the savings in time from having all the connections pre-wired is well worth the cost in the long run.

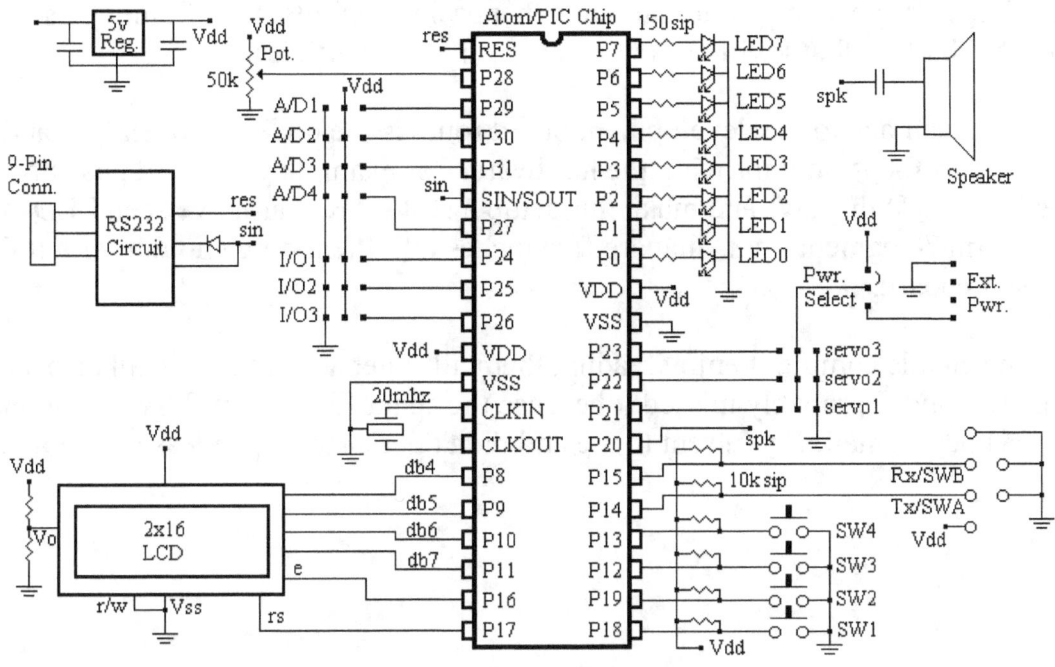

Figure 3

Summary

It's little projects like this that build up over time. Once you have a few development boards lying around you'll find yourself writing quick snippets of code to test something out. Soon you have a library of sample code for creating all kinds of little gadgets. I've found it's the hardware that takes so long to get it right. This is why I've developed all my breadboard modules that you can purchase at www.beginnerelectronics.com. I use them to quickly plug and play ideas and then use that knowledge in future projects. They don't eliminate the hardware but reduce it down to a few jumper wires. When you want to make something quick just to prove that it will work though you cannot beat the simplicity of a pre-built development board.

The Atom also adds a lot of simplicity since I don't need to use a separate programmer to get the code into the PIC. Just a simple serial connection will work. I've even tested the BasicBoard with a few RS232 to USB adapters and most worked well. You still need to supply separate power though. Look for that option when searching for a development board. Serial ports are getting harder to find especially on laptops. If you can find a board with USB to RS232 built in that is great. There are new chips out that make that easier to implement. Some PICs even have USB built in that I will be writing about in future articles.

Another advantage to the development board route is the code is proven first and then you can focus on improving the hardware. For example you could possibly drive large LED displays and make this setup a real scoreboard. I've seen LED arrays form 7-segment digits that are five inches tall. Plenty big enough for a full scale scoreboard.

Keep the emails coming. I enjoy reading them all. I get a lot of junk mail also so if I don't respond, I probably missed it because the spam filter got it. I try to respond to everybody's email. If you sent to me and didn't get a response please try again.

March 2007 - In-Circuit Serial Programming

Reader feedback is so important to me because without it I'm writing in a vacuum. Just because I think a particular topic is interesting, doesn't mean the readers will. I've also seen a lot of emails from new visitors to the column and they want to see some of the topics I've already covered. One topic seems to come up often in reader feedback and that is the subject of In-Circuit Serial Programming (ICSP). This can be a very useful feature so I thought I would cover it in detail this month.

The advantage to ICSP is the ability to program the PIC in-circuit. Now this may sound familiar since this is the same feature I've promoted about using a bootloader in a previous column. ICSP and Bootloaders are similar in function but the difference is the type of hardware support required. Bootloaders use an RS232 or USB interface circuit between the PIC and a PC serial port connection. ICSP uses a PIC hardware programmer between the PIC and the PC. In fact, many PIC programmers use ICSP to program the PIC even if you are putting the PIC in a socket. My EZPIC programmer with the ic-prog.com software does this. The steps you need to make a PIC design in-circuit programmable so you don't have to remove the PIC are actually quite easy. The best part is it will work with almost all PICs while a bootloader typically won't work with the smaller PICs unless you make your own custom version for each particular PIC family.

The biggest hang-up I've run into with readers who've tried ICSP is the serial communication signal gets affected by the circuitry connected to the PIC. For example, to program a PIC in-circuit using ICSP you need five connections; 5v (Vdd pin), Ground (Vss pin), Vpp (MCLR pin), Data (PGD pin) and Clock (PGC pin). Many PIC programmers have these pins available on some kind of header so the hardest part is making a conversion cable from your programmer to your circuit or circuit board. If the Clock or Data pins are not able to send the correct signal, the PIC will not program properly and you will get a verify error. This can easily be corrected.

Before I get into that I wanted to re-introduce one of the USB programmers I mentioned in the October 2006 article which is the PICkit 2 programmer designed by Microchip. This programmer comes as a complete package for under $50 from microchipdirect.com and it is designed to easily plug into a 6-pin ICSP header. Five of those six header pins are the five ICSP connections I mention above. Figure 1 shows the PICkit2 package. It's small and USB powered so this makes a great programmer to place between the PC and your circuit for everything I'm about to talk about. I'll also show you a great ICSP feature of this programmer but let's first cover the hardware requirements for ICSP.

Figure 1

Hardware for ICSP
The schematic in Figure 2 shows the five ICSP connections and all the possible connection issues to watch out for in your design. Because of the way the ICSP feature works, you don't want to add any capacitance to the programming connections since this can delay the signals. Even the capacitance on the Vdd line should be monitored per the PIC programming specification. The PIC programmer actually cycles the Vdd line off and on while sending the Vpp signal to the MCLR pin. This is done to put the PIC in programming mode. If there is too much capacitance, it may slow the signal down and not meet the programming specs. You can get the programming specs for any PIC at the Microchip.com website.

You also don't want to load down the clock or data signal. The components that are crossed out show what not to include in your circuit. The diodes on the Data and Clock lines are a mistake because there is two-way communication when programming and verifying the part. These are pretty easy to see why they should not be included in your design.

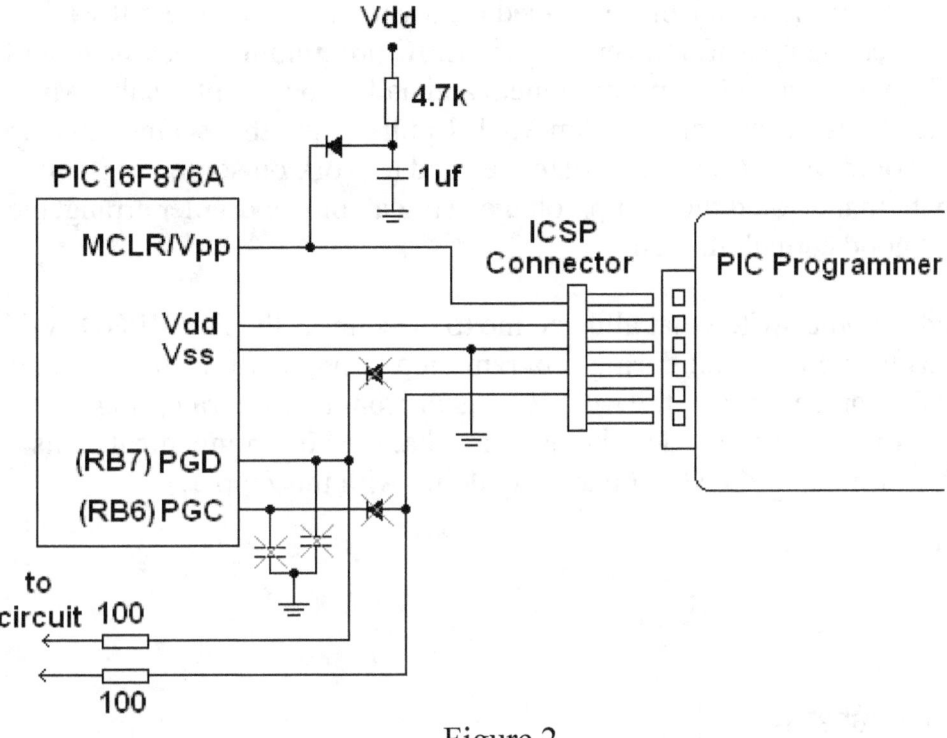

Figure 2

What isn't quite so clear is the diode between the MCLR reset circuit and the MCLR/Vpp pin. This is recommended because the PIC programmer sends a high voltage signal to the Vpp line of around 12v -13.5v for a period of time. You don't want that signal feeding into your Vdd regulator. This is actually just a safety precaution though because the current is small and the MCLR pull-up resistor will knock it down to prevent any damage. I use a 4.7k but a 1k will work fine.

Another recommendation which is often missed is the series resistors on the PGD and PGC lines between the PIC and the rest of the circuit. These isolate your circuit from the PGD and PGC signals so your circuit doesn't load down the PIC programmer. This is usually where people have a problem with ICSP. One hundred ohm resistors should not affect your circuit function but it should be plenty of resistance to isolate the programmer.

Another approach to ICSP isolation is to add a switch to your circuit. This is the way I handled it on the original version of my Zipper board. A long time ago when

I was just getting started with PicBasic, I wanted a simple Basic Stamp like module to program. The Zipper was the result. I didn't know about the clock and data series resistance idea or the MCLR diode suggestion because I didn't read all of the data sheet information. Before I added the switch I could not get ICSP to work properly. I used the microEngineering Labs EPIC programmer back then but I tested ICSP with several PIC programmers and had inconsistent results. My solution was to use a smaller 100 ohm MCLR pull-up and that seemed to help. The smaller 100 ohm MCLR pull-up resistor seemed to work on some programmers I tested it with that created their Vpp voltage using a voltage divider arrangement. It still wasn't good enough though.

I then tried a 4 pole switch that allowed me to disconnect the PGC, PGD, Vdd and MCLR pins from my circuit during programming to prevent it from loading down clock and data and eliminate the MCLR resistor from the programming connection. Figure 3 shows the schematic for that type of arrangement. This worked very well and it's what I ended up doing with the Zipper.

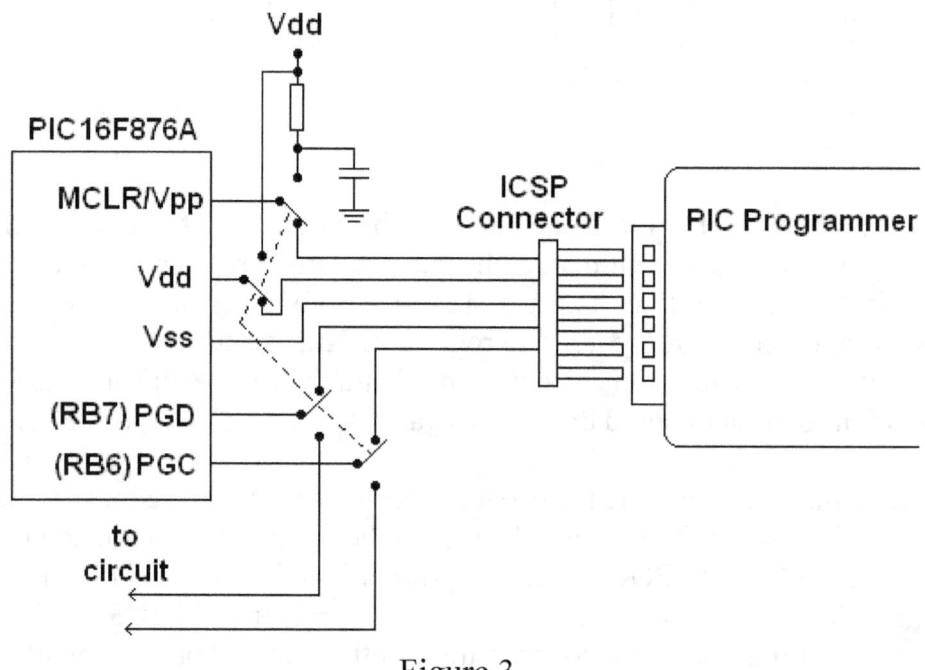

Figure 3

Figure 4 shows the finished original Zipper design with the ICSP header setup for the microEngineering Labs EPIC PIC Programmer which was very popular back then. It's still one of the best parallel port programmers available. To use the

Zipper I would just slide the switch to program position to download code and then slide it back to run the code. Eventually I replaced the EPIC header with a 6-pin SIP header to match the Olimex PG1 programmer that is a surface mount version similar to my EZPIC programmer. This way I could include a PIC programmer with the module for a lower price. I'm slowly phasing out that design as I use the Ultimate OEM bootloaders more than anything else now but this is another option for successful ICSP.

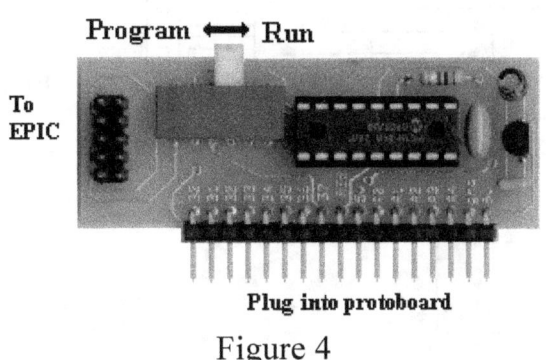

Figure 4

12F675 Trick

Now this next section should be worth the price of your subscription. While playing with the PICkit2 programmer from Microchip, I actually read the manual. No really, I did. And it details that the 6th pin on the PICkit2 ICSP connector, also known as the AUX pin, can be used to regenerate the internal oscillator calibration OSCCAL value that is set at the factory and placed in the last location of program memory. Remember I talked about how that value could be erased if you erase the whole PIC 12F675 chip back in the January article? Well, if you tie the AUX pin to the GP4/T1G pin of the 12F675, as seen in Figure 5, the PICkit2 software will generate a new OSCCAL value and put it at the last location of memory just like it wasn't erased.

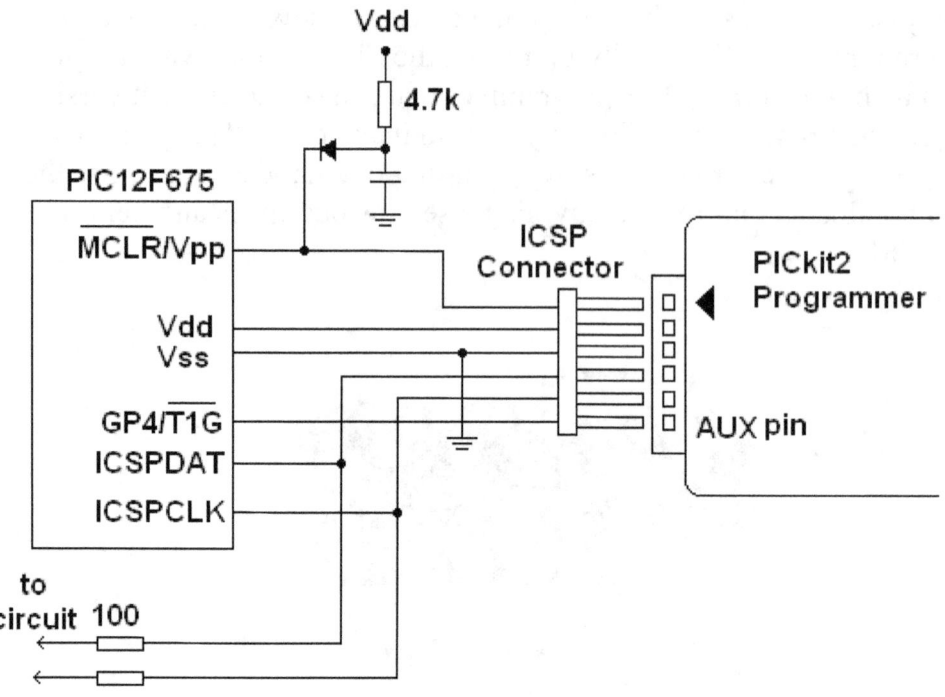

Figure 5

I had to try it out and it worked though it didn't get the exact same value every time but it was very close. Not sure how it works but it's a great option. Notice that the Data and Clock lines are called ICSPDAT and ICSPCLK on the 12F675 parts. The labeling changes a little from part to part in the data sheets for some reason.

Quick & Dirty Board

I recently wanted to try out one of the new 16F887 PICs that I just got from Mouser.com. These are the next generation of PICs that take the 16F877A to a new level. Plus they are cheaper than the 16F877A. Microchip also offers a 16F886 to upgrade the 16F876A that I like to use. I ordered some of those from mouser.com also but haven't received them yet. The latest version of PicBasic Pro, which is 2.47, adds support for these parts. My problem is my EZPIC programmer relies on the ic-prog.com software and it hasn't been updated for this part yet. The PICkit2 does support the 16F88x parts but the board that comes with it doesn't have a 40 pin or 28 pin socket. Therefore I decided to use this ICSP method to create a quick and dirty 40 pin programming socket for the PICkit2. Figure 6 shows the result.

Figure 6

It's not pretty but it worked. I even added a couple LEDs so I could test it with a few flash LED programs. The 16F88x parts have internal oscillators so I didn't need to add an external resonator. This was an upgrade from the 877A. I also could tie the MCLR pin high internally with a configuration setting so I didn't need the external resistor. The PICkit2 can also power the circuit through the 6-pin ICSP connector Vdd and Vss connections but it is limited. The USB port can only supply 100ma and the PICkit2 draws some current. If you need less than 50 ma you should be fine powering from the PICkit2 but don't quote me on that, I'm just speculating. Since this was so simple I let the PICkit2 power this guy while it flashed the LEDs. The project worked great so now I'm thinking about laying out several different plug-in boards for my home lab. Being able to simply plug in the PICkit2 to both program and power the board is a nice feature.

ICSP Pin-Out

As a final note I thought I would give you the ICSP pins for the various PIC DIP packages. Table 1 below summarizes it for you. This should save you the trouble of looking through all those data sheets.

ICSP Connection	40 pin DIP	28 pin DIP	20 pin DIP	18 pin DIP	14 pin DIP	8 pin DIP
Vpp\MCLR	1	1	4	4	4	4
Vdd	11 & 32	20	1	14	1	1
Vss	12 & 31	8 & 19	20	5	14	8
Data	40	28	19	13	13	7
Clock	39	27	18	12	12	6

Table 1

Conclusion

Hopefully I've explained this well enough that you can plan your next PIC board layout around ICSP. One of the advantages to using the bootloader in my Ultimate OEM module was the ability to use a simple serial cable as the connection device and the ability to run the MCStudio Plus in-circuit debugger that I talked about in a previous article. With the small size of this PICkit2 and also adding to the fact that the PICkit2's cousin, the PICkit2 Debug Express can do in-circuit debugging through the same 6-pin ICSP connector, I'm forced to rethink my development module layout. I have a friend who's been working on his own development board designs based on this new PICkit2 setup. I'll see what he's up to and maybe report that out here in another article. The PICkit2 Debug Express only supports a few PICs right now but I expect that to grow.

Send me your emails on future topics you'd like to see. I know several people will read this and say, he listened, he actually wrote what I asked! If nobody reads my articles, N&V will tell me to hit the road so why would I do anything else. I just have to keep it interesting and progressing as the reader's knowledge expands about PICs. That's what I'm trying to do so please email me your feedback. I read it all. Send it to: chuck@elproducts.com. You can always visit my website as well at www.elproducts.com. See you next month.

April 2007 - PIC vs PIC Speed Test

If you look through Nuts & Volts or surf the internet you'll see lots of different Microchip PIC based development chips and modules offering all kinds of different features. Some of the more popular ones are the PicAxe, Basic Stamp, Basic Atom, my own Ultimate OEM and even my BasicBoard just to name a few. Which is the best platform to start with or what is the best bang for the buck are topics of questions I often get in email. I also get readers asking how these compare to programming with a Basic compiler like PicBasic Pro. In this months column I'll compare them all with a common program to see how well they do against each other.

The emails I get often come from readers of this column who are just getting started programming and have also read the Basic Stamp column or read about the PicAxe or read about the Basic Atom and wonder what the difference between all these are. Because of all these choices they begin to doubt their own instincts. In the body of the emails they might ask if the Basic Stamp is really a Microchip PIC based module since they read something about Scenix. Or they read about the PicAxe and wonder why it doesn't need a hardware programmer. They read another book or article and it tells them only a "real compiler" is the way to go if they plan to take their design into production. They begin to question the advantage of PicBasic Pro over the single chip approach such as PicAxe. In most cases the limiting factor comes down to cost. Let's face it, if money wasn't and issue you could own them all and evaluate them for yourself but that's not often possible. Fortunately I've collected several of these systems over the years because I to was curious and decided to run my own tests.

Back in the early days of computers there was a speed test program I could run at the DOS prompt that would give me a performance rating on the PC. By comparing this rating I could determine if one PC was really better than the next. I thought this would be a great way to compare these Microchip PIC options. You see by measuring speed you are also measuring code size. If it takes a long time on one module to process the same function than it does on another module then it must use more PIC commands. No matter how the code is written it eventually gets down to binary 1's and 0's. I decided to reproduce that test to compare these various PIC based setups. I'm chose five popular hobbyist approaches; Basic Stamp 2, PicAxe 28x, Basic Atom 28, PicBasic and PicBasic Pro. I'll also compare a popular "C" compiler for this test to see if programming in C is that much faster than Basic. For those that aren't familiar with these choices I'll begin by briefly describing them.

PicAxe

I have to admit when these first came out I wasn't impressed. They based their operation on the original Basic Stamp 1 (BS1) PBasic command set which is limited. The original BS1 was built around a PIC16C56 with an external EEPROM memory for storing the commands. The PicAxe (figure 1) reduced those two chips into a single chip with software. I didn't see that as a big advantage but on the other hand I was impressed with the free PicAxe flow chart style GUI programming software. This made it easier for kids to get involved. PicAxe also offers the ability to program directly in PBasic commands just like the BS1. I began to re-think my first impression. The PicAxe has since grown to include a whole variety of options and new commands beyond the BS1. This all comes at a very low price as you can get PicAxe chips for only a few dollars and also very low cost starter kits.

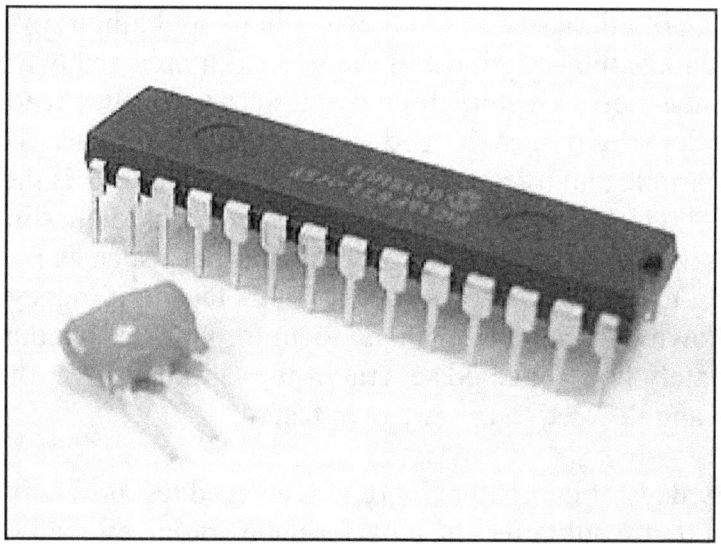

Figure 1

Every PicAxe is a Microchip PIC with a custom self programming software program installed. At first I thought it was a bootloader but since some of the PICs they use don't offer self-write memory I figured this wasn't the case. The small 8-pin version is based on the a PIC12F629 and the 28x, that I'll use in my test, is based on a PIC16F873. The other versions are PICs but I'm not sure which ones since I cannot find details on them and I only have these two I've mentioned in my lab. It's clear though that they work similar in that they program themselves through a serial connection. The BS1 operated the same way and stored command tokens in external EEPROM. My guess is the PicAxe stores command tokens in internal EEPROM and then retrieves each command one at a time and processes it. Therefore when you program a PicAxe you are passing tokens to it that get stored in EEPROM by the resident software already on the chip.

Each PicAxe chip has a limited number of total commands defined with the maximum being 600 commands. This seems to relatively match the EEPROM size of the particular package size. The 8-pin has a smaller EEPROM and smaller command limit and the larger PICs have more EEPROM and larger command limit. What I questioned most was the speed. The PicAxe 28x runs with an external 4 Mhz resonator and I suspect the 8-pin PicAxe and the others run on the internal 4 Mhz resonator that these PICs have. I know the BS1 also ran at 4 Mhz and that PicBasic compiled BS1 programs ran faster than they did on the BS1 so I was very curious how fast the PicAxe ran. Let's cover the other players before that.

Figure 2

Basic Stamp 2
The Basic Stamp 2 (BS2), shown in figure 2, was a great module when it was first released. It was one of the first PIC based modules for those that wanted to use the easy to understand Basic language to program PICs. The BS2 also has a free programming software that is command based so all you need is a BS2 module or build one from the BS2 interpreter chip and EEPROM. The interpreter chip is more money than a PicAxe chip but I suspected it would run faster than the PicAxe. The BS2 advertises 4000 instructions per second while the BS1 is only 2000 instructions per second. The BS2 also runs on a 20 Mhz resonator vs. the PicAxe and BS1 4 Mhz. This often confused me since the BS2 ran an oscillator 5 times faster than the BS1 but only advertised instruction speed of 2 times faster.

The BS2 is based on a PIC 16C57 and also stores command tokens in external EEPROM. The access time to read and write to external EEPROM is slower than writing to internal EEPROM on a PIC so that will affect processing speed during my test. I began to wonder if the PicAxe with internal EEPROM would be as fast or faster than the BS2 even though the BS2 has the faster oscillator.

Basic Atom

The Basic Atom (figure 3) takes the PicAxe/Basic Stamp programming style a little further in that it pseudo compiles or combines the token style commands with the processing routines and stores then in a binary .hex file. This binary .hex file gets programmed into the PICs program memory. This way the program is retrieving information from within program memory rather than EEPROM memory. This is much faster than EEPROM access but uses a lot of program memory initially when it downloads all this to the PIC. The Atom is based on the PIC16F876A and 16F877A both of which have the capability to self-write to program memory from within the software. This is how a bootloader works and a bootloader is how the Atom programs itself through a PC serial connection. This is why an Atom doesn't need a separate PIC programmer, similar to PicAxe and BS2, to download the .hex file it creates.

Figure 3

PicBasic

PicBasic (figure 4) is the original Basic compiler from microEngineering Labs and it takes BS1 programs and directly converts them into an assembly language file that gets assembled into a true binary .hex file. This .hex file can then be programmed into a blank PIC with a hardware PIC programmer. This is the compiler that got me started programming blank PICs in Basic. I had reached a limit with the BS1 and had moved on to the BS2 when PicBasic arrived on the scene. Programming in assembly language just was too time consuming for me. I was so amazed at how easy it was to program off the shelf PICs which were a ton cheaper than Basic Stamp modules or interpreter chips. The PicAxe and Atom weren't around when the PicBasic compiler was introduced so those weren't an option.

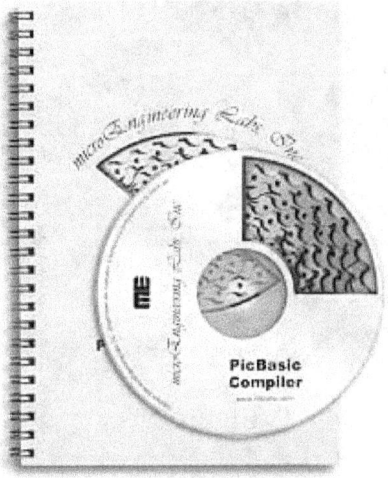

Figure 4

The PicBasic compiler added a few commands to the BS1 PBasic command set that allowed it to access registers in the PIC and it also allowed more RAM space to be used for variables.This was a huge advantage over BS1 and BS2. PicBasic was the best thing going until the PicBasic Pro compiler was released. PicBasic is limited to 2k of PIC program memory so even if you have an 8k PIC you would have to do some special coding to get beyond 2k. That was still a huge improvement over BS1 and BS2. I thought is was interesting to add PicBasic to my performance test since I haven't used it in a while and haven't even talked about it in this column series. I do include both PicBasic and PicBasic Pro in my book "Programming PIC Microcontrollers with PicBasic" though.

At $99.95 it's a pretty good alternative to programming with PicAxe or BS1 since the commands are the same and you can use it with lower cost blank PICs. PicBasic is also written to run programs at 4 Mhz. You can run the PIC faster but some of the time critical commands will be affected. For example a PAUSE 1000 will delay one second at 4 Mhz but run the PIC at 20 Mhz and that command will only take 200 milliseconds. Some people use this method to get higher baud rates out of PicBasic's SEROUT command. For the performance test I ran it at 4 Mhz.

Figure 5

PicBasic Pro

The PicBasic Pro compiler (figure 5) has been talked about a lot in this column. It added so many features to the PicBasic compiler, it clearly was in a class by itself when it was released and is still one of the more popular compilers on the market. The PicBasic Pro compiler produces an assembly file that gets assembled into a binary .hex file so it can program just about any 8-bit PIC out there. I noticed right away when I switched from PicBasic to PicBasic Pro that it greatly reduced the memory size of my programs in the exact same PIC.

PicBasic Pro isn't limited to 2k like PicBasic so it's number of commands is only limited by the program memory space of the PIC you are using. If you use a PIC18F6722 you get 128k bytes of program space. I rarely use more than 8k so that's one huge advantage of PicBasic Pro. How much better it is at speed was really the question I wanted to know. PicBasic Pro can be setup to run at various speeds with a simple DEFINE insert so this allowed me to run the same program at both 4 Mhz and 20 Mhz.

Speed Test

And now for the moment we've all been waiting for, the speed/performance test. I wanted to create a simple program that could be used across all these different parts with very little or no modification. I knew I wanted it to process several commands in a loop and also perform some math. I also wanted to light an LED so I could see it working and be able to put an oscilloscope on the LED to measure the time it took to change state. This change in state delay reflects the processing time so by recording that time I could compare these parts running the exact same Basic code. The final program is below:

```
Main:
high 1              'Set Port 1 High to Light LED
For x = 1 to 255    'Loop 255 times in for-next loop
y=y+1               'Perform simple math
next                'End for-next loop
low 1               'Clear port 1 to turn off LED
pause 10            'Stay low for 10 milliseconds
goto main
```

On each part I had to create the x and y variables in slightly different ways. In PicAxe and PicBasic I used:

```
Symbol x = B0
Symbol y = B1
```

On the others I used the VAR directive:

```
x var byte
y var byte
```

These variable declarations didn't affect the speed since they are a one time function. I was really surprised how easy it was to make one program fit all these different platforms which just shows how common the Basic language for PICs is. Each processor reported a different amount program space and I was a little confused on how to measure some of them. I could not figure out how much memory I used in the BS2 but I'm sure that was because of my inexperience with the Stamp programming environment. For the most part though, program size affected speed. The slower the part, the more memory it used.

Results

The picture and chart below (figure 6) show the results and what a difference there is. The PicAxe came in the slowest and PicBasic Pro at 20 Mhz was the fastest. I'm probably not doing it justice by just saying fastest as it was over 1000 times faster to run the program in a PIC16F876A with PicBasic Pro at 20 Mhz than it was to run it in a PicAxe 28x. That was amazing how different it was. You can also see how the move from external EEPROM to internal program memory helped Atom beat Stamp by 4 to 1. I was even surprised to see PicBasic Pro at 4 Mhz beat PicBasic by such a large margin. The Pro program was about 1/3 the size of the PicBasic program also.

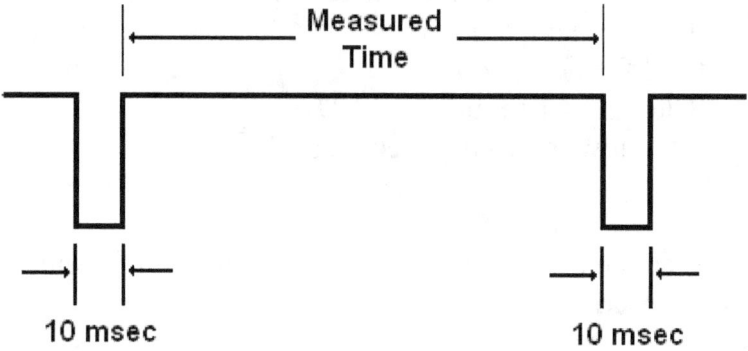

Figure 6

PIC Option	Measured Time
PicAxe 28x @ 4 Mhz	368 msec
Basic Stamp 2 @ 20 Mhz	278 msec
Basic Atom @ 20 Mhz	68 msec
PicBasic in PIC16F876A @ 4 Mhz	28 msec
PicBasic Pro in PIC16F876A @ 4 Mhz	1.76 msec
PicBasic Pro in PIC16F876A @ 20 Mhz	0.350 msec

C vs PicBasic Pro

Now this is a topic I have wondered about for years. I get emails from C programmers asking me to stop using that lousy interpreted Basic code and program in a real language such as C. I then explain that PicBasic Pro is really compiled just like C and in many cases will produce code the same size as a C language compiler. I'm not here to say PicBasic Pro is better than C or C is better than PicBasic Pro. All I know is compiled Basic language for PICs gets very little respect and C gets too much if you ask me.

I decided to write this same program in C as close as possible to the Basic version using the HiTech PICC-Lite compiler from htsoft.com. This is a popular PIC C compiler and I like it a lot. I will not claim to be a great C programmer but the results of this test were very interesting. The PICC-Lite compiler is a free download and it works with the PIC16F877A so that is the chip I used since it's essentially the same PIC as a PIC16F876A with more I/O. The speed test code in C is below:

```c
unsigned int counter;    // Create delay loop variable with max range of //
                         0 to 65535

void Pause( unsigned short usvalue );//Establish pause routine function
void msecbase( void );        //Establish millisecond base function

main()
{
PORTB = 0;                    //Clear PortC port
TRISB = 0;                    //All PortC I/O outputs


while(1==1)                   //loop forever
{
    unsigned int z, y;
    RB0 = 1;                  // Turn on RC0 LED
    for(z=0; z<256; z=z+1)
    {
    y=y+1;
    }
    RB0 = 0;                  // Turn off RC0 LED
    Pause(10);                // Pause 10 msec


}       //End while
}       //end main
//****************************************************
//pause - multiple millisecond delay routine
//****************************************************

void Pause( unsigned short usvalue )
{
    unsigned short x;

    for (x=0; x<=usvalue; x++)//Loop through a delay equal to usvalue
        {                     // in milliseconds.
        msecbase();       //Jump to millisec delay routine
        }
}


//****************************************************
//msecbase - 1 msec pause routine
//****************************************************

void msecbase(void)
{
    OPTION = 0b00000001;      //Set prescaler to TMR0 1:4
    TMR0 = 0xd;               //Preset TMR0 to overflow on 250 counts
    while(!T0IF);             //Stay until TMR0 overflow flag equals 1
    T0IF = 0;                 //Clear the TMR0 overflow flag
```

47

}

This will look very confusing to the beginner and Basic programmer since C is very cryptic the first time you look at it. If you compare the core section of both versions though you see a lot of similarities as shown in the block below:

C Code	PicBasic Pro Code
``while(1==1) //loop forever
{
       unsigned int z, y;
       RB0 = 1; // Turn on RC0 LED
       for(z=0; z<256; z=z+1)
       {
       y=y+1;
       }
       RB0 = 0; //Turn off RC0 LED
       Pause(10);// Pause 10 msec

}   //End while``` | Main:
high 1   'Set Port 1 High to Light LED
For x = 1 to 255
y=y+1   ' Perform simple math
next
low 1   'Clear port 1 to turn off LED
pause 10   'Stay low for 10 milliseconds

goto main |

I then compiled the C code and ran it at 4 Mhz to compare it to the rest of the pack. The results below show how PicBasic Pro at 4 Mhz compares to PICC-Lite at 4 Mhz.

Compiler	Measured Time
PICC-Lite in PIC16F877A @ 4 Mhz	3.6 msec
PicBasic Pro in PIC16F876A @ 4 Mhz	1.76 msec

I have to admit I was shocked by these results and I'm sure I'll get a lot of email on this one about how I screwed up somewhere but I tested it on several PICs after this and got the same results. PicBasic Pro was faster than C in this simple application. To further understand I reviewed the .hex files and saw that PicBasic Pro compressed the code into a single block of space that took up 60 words of program memory. The PICC-Lite version broke up the code into two blocks; one at the top of program memory and one much further down. This told me that PICC-Lite had produced code that possibly jumped around more than PicBasic Pro. If your assembly code has more calls or goto's in it then it will take more time to process. The PICC-Lite file also took 68 words of space. This was larger than PicBasic Pro.

Now this is the freeware version of PICC and it doesn't have all the optimization layers in place that help compact the code so there are improvements that can be made. Compacting can sometimes add more calls so I'm not sure that helps much. Inserting assembly into C helps speed things up but then again you can do that in

PicBasic Pro. All in all I was impressed at how well PicBasic Pro performed against PICC-Lite in this test. If you want to become a professional software engineer then C is still the most desired language to know right now. This tells me that PicBasic Pro can compete well in applications where you may need to rely on that simple Basic language to get a product out the door quickly. I've help many non-software engineers get products out the door using PicBasic Pro so this just confirms in my mind that this is not a compromise. I've heard of PicBasic Pro projects running on the shuttle so it must work for NASA as well.

Conclusion
Based on this brief study I've concluded that no matter how you get started you end up getting what you pay for. If you only have a few bucks then PicAxe is a great place to start and you will learn the language that can later be used with PicBasic Pro. Basic Atom and Stamp offer a smooth upgrade path to PicAxe and PicBasic/PicBasic Pro offer the ultimate option for programming low cost PICs.

Purchasing a decent PIC programmer and PicBasic Pro will cost a few hundred dollars but divide that up amongst all the projects you create with it and it will eventually cost the same as PicAxe. And once you start reading sensors or trying to process large amounts of data then speed becomes a priceless advantage. I still do a lot of work with Atom because of the great built in in-circuit debugger (ICD) and I like some of the commands it offers that PicBasic Pro doesn't but for the serious jobs I tend to use PicBasic Pro. Now I have a clear reason why.

I hope you learned something here that will help you decide your best route to get started. If you have any questions or comments send them to me at chuck@elproducts.com or visit my website at www.elproducts.com for all your "Programming PICs in Basic" needs. See you next month.

Special Note:
There is a correction to this speed test article that was published in the June 2007 article. Please refer to it for the complete details.

May 2007 - 8/14/20-Pin PIC® Microcontroller Family

Programming PIC® microcontrollers (MCUs) has been my hobby and sideline for many years. In fact I started back when there were only five PIC MCUs to choose from. Now there are 100's of different PIC MCUs. I've helped 1000's get started programming PIC MCUs through my books, articles and development boards and have enjoyed every minute of it. Outside of this activity, my professional electronics career (that pays the larger portion of my bills) has always been in the automotive electronics industry, and I've learned a lot about designing electronics in this harsh environment. Change is inevitable and I recently was offered the chance to join Microchip as a Field Applications Engineer, so I made the move. This should give me even more information to draw from for these articles, but my focus for this column is and will always be to help beginners, hobbyists and anybody else learn how to get started programming PIC MCUs.

Before I joined Microchip, I hadn't used the 14-pin or 20-pin PIC MCUs at all. In fact, I think I only had one in my storage cabinets, which are full of PIC MCUs. The 14-pin and 20-pin PIC MCUs are cousins to the 8-pin PIC MCUs that help make up the full line of low pin count devices. They all share the same Vdd, Vss, Data Clock and MCLR connections that are needed for programming, which I only realized during the development of my EZPIC programmer. I originally had an 8-pin socket, and then I changed it to a 14-pin socket when I realized they could share. After getting feedback from a customer who had successfully programmed a 20-pin PIC MCU by putting it in the 14-pin socket, I modified the design to replace the 14-pin with a 20-pin socket. I'm just in the process of getting those boards manufactured, because I had to wait until the previous version was sold out.

Common Pin Out

I did the design work on my EZPIC programmer before joining Microchip, but never fully realized how handy having those common pin outs was until I started playing with them. Because I had easy access to some of these parts at Microchip, I began using them with Microchip's PICkit™ 2 programmer, which I've talked about here before. The PICkit 2 starter kit comes with a 20-pin socketed development board. I began by programming various 8-, 14- and 20-pin PIC MCUs to better understand all of the features these parts offered. I soon appreciated how I could build sample programs using the various parts 8-, 14- and

20-pin PIC MCUs without having to change the development board. This was the result of the common pin out between these parts. Figure 1 shows the common pin layout.

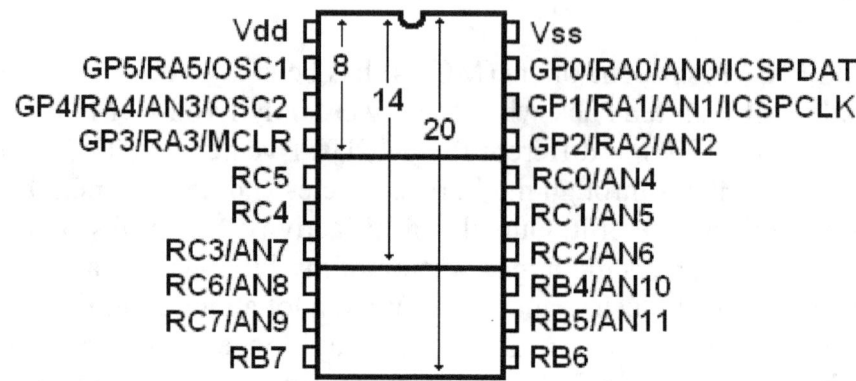

Figure 1: The common pin outs of 8-, 14- and 20-pin PIC MCUs

Notice how the 20-pin parts have up to 12 A/D ports. There is a lot packed into these parts, so I left off some of the pin descriptions–such as the comparator pins and the T1G pin designation for the Timer 1 gate. This a relatively new feature on PIC MCUs that allows you to turn the internal clock feed to Timer1 on or off, based on an external signal. You can set it to allow the internal clock feed to run on a high pulse and turn off on a low pulse. This way, you can use Timer1 to easily measure pulse width in the background while your program is doing something else. This will definitely be a feature I'll talk about in a future article.

My January 2007 article on using the PIC12F675 as a 555 replacement was popular, based on email feedback. Apparently, a lot of people like those little 8-pin devices. Now I realize that starting with an 8-pin PIC MCU lays the groundwork for a smooth upgrade path. If you plan ahead and lay out your circuit board for a 20-pin PIC MCU, you can just drop in a 14-pin or 8-pin PIC MCU in the same socket without issue. You can use the extra I/O on the 20-pin part for functions you will only use occasionally. All of these parts offer the internal oscillator plus the internal MCLR option, so all you really need to hook up is Vdd (2.0-5.5v) and Vss (ground) to bring the PIC MCU to life.

Extra Features

Some of you may be reading this and saying, "Duh, where have you been? We've known about these parts." Sorry if I'm telling you something you already knew, but I was surprised to find how many advanced features these parts have. One of them is the Enhanced Capture and Compare PWM (ECCP) peripheral that is great for designing a motor-control application. Figure 2 shows a sample setup direct from the data sheet. My layout in Figure 1 doesn't show the P1A thru P1D pins, but you can easily find them on the data sheet.

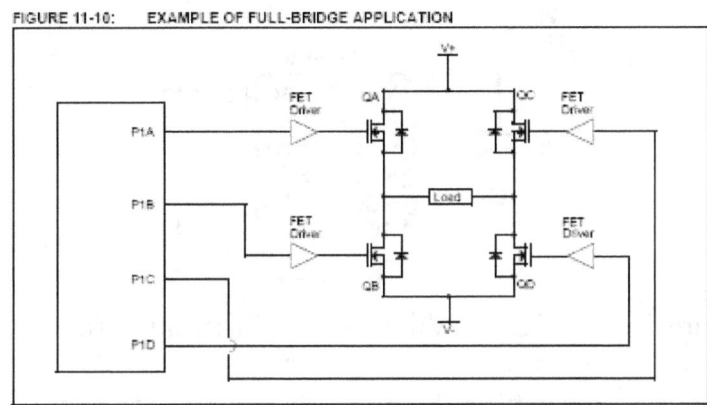

FIGURE 11-10:   EXAMPLE OF FULL-BRIDGE APPLICATION

Figure 2: An example setup for the Enhanced Capture and Compare PWM

I hope to cover more on this peripheral in a future article, as I know motor control interests a lot of people. The nice thing about this peripheral is it handles all the timing and even the dead-time delay required, so you don't ever have two FETs on at the same time causing a short. This can happen when one FET is shutting down slowly while the other starts up quickly, causing the temporary short.

These parts also have internal comparators and even share some of the same A/D connections, shown as all the ANx pins in the layout of Figure 1. I really have to look at these parts for future home projects, because they have some of the latest and greatest PIC MCU features. They even have a LIN peripheral in some of the parts, which is a communication bus that is growing in popularity within the automotive world. Again, maybe you knew about all this but I just wanted to show that even an old-timer like me can work with PIC MCUs for years and still discover new stuff.

High-Voltage (HV) PIC MCUs

Through this discovery phase, I also was introduced to a couple of new 8/14-pin family members—the PIC12HV615 and PIC16HV616, which are newly released.

These are Flash-based PIC MCUs, but they have a built-in shunt regulator (which is a fancy term for a zener diode). This means they can run off of a higher voltage without needing an external regulator. Figure 3 shows a schematic for a PIC16HV616 with the shunt-regulator series resistance and capacitance in place. I'm not showing the MCLR pull-up or the oscillator, since these would be set to internal operation.

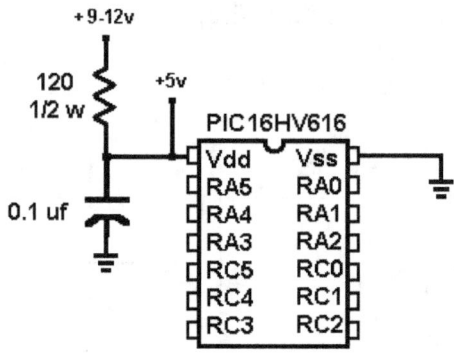

Figure 3: The PIC16HV616 with built-in shunt regulator

The key to using this part is setting the proper series resistance and capacitance. There are three formulas for calculating those values, depending on the current and voltage range you need to work with. I've found that you either have a large voltage range or a large current range, but not both at the same time. For example, I chose to use a voltage input of 9 to 12 volts. This limits how much current range I can have. I limited my current variation to 20-25 ma. If I go above or below this, the shunt regulator will be out if its design limits. Most people can set their design to a fixed voltage input, which gives you a greater current range to work with. The equations are listed below as Equation 1, 2 and 6, which I pulled from the Microchip application note AN1035 and you can download from www.microchip.com.

EQUATION 1:

$$R_{MAX} = \frac{(VU\_MIN - 5.0)}{1.05 \cdot (ILOAD\_MAX - 4\ mA)}$$

EQUATION 2:

$$R_{MIN} = \frac{(VU\_MAX - 5.0)}{0.95 \cdot (ILOAD\_MIN + 50\ mA)}$$

EQUATION 6:

$$C_{MAX} = - \frac{42 \, \text{MS}}{R_{SER} \cdot \ln \left(\frac{2.1}{5.0} \right)}$$

Equation 1 gives us the upper limit of the series resistance. Using the lower end of our voltage range and the upper end of our current range, we get the following:

Rmax = (9v – 5v) / 1.05 (25ma + 4 ma)
Rmax = 131 ohms

Equation 2 gives us the lower limit of the series resistance.

Rmin = (12v – 5v) / 0.95 (20ma + 50 ma)
Rmin = 105 ohms

With this, I select a value of 120 ohms that falls within the two limits. But, I need to calculate the power rating of the resistor. Since the top of the resistor can see a maximum of 12V and the bottom will see the regulated 5V, the power is found with the equation:

Rpwr = (12v – 5v)^2 / 120 ohms
Rpwr = 0.4 watts

Therefore, I'll use a ½ watt resistor.

Now I have to calculate the capacitor. The data sheet states the capacitor needs to be larger than 0.047 uf for noise suppression, and less than the calculated capacitance of equation 6. So, putting the values in the equation give us the following:

Cmax = 42 / [120 * ln (2.1 / 5.0)]
Cmax = 0.4 uf

Therefore, I choose a value of 0.1 uf—since it's greater than 0.047 uf and less than 0.4 uf. We now have our PIC MCU regulator ready to go. One thing to note with this type of setup is the amount of current a shunt regulator draws without anything happening. The shunt regulator will draw 4 ma, even if the PIC MCU is in sleep mode and nothing else is connected. This is not an ideal setup for battery-operated devices that need long life. For those devices, you could use the PIC16F616 version of this same part, with an external low quiescent current draw regulator. The HV parts are really handy, though, for applications that need a small component count. I've seen motors with the HV part built into the back of the motor. This may be a great way to make a serially controlled motor—similar to a serially controlled LCD module. (That also may be the focus of a future article.)

What I like most about these parts is the ability to run them off the same voltage source as the outputs you may be trying to drive. For example, back in the November 2006 column I showed some hardware interfaces to the PIC MCU. One of them showed how to drive some high-voltage devices using a bipolar transistor, as seen in Figure 4.

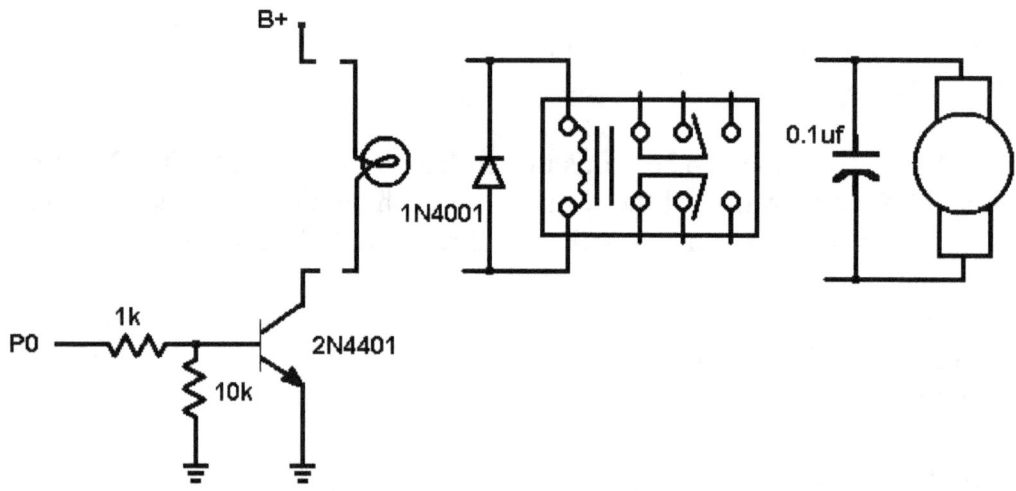

Figure 4: A configuration for driving high-voltage devices

I don't specify a voltage here, but instead just state B+. What if that was 12 volts, instead? Using the resistor and capacitor values I calculated above, I could control one of these outputs while also powering the HV PIC MCU from the same supply. Now, this will require some better filtering to prevent the large voltage line from dipping quickly when the load kicks in, possibly resetting the PIC MCU. But, that may be easily solved if the load is fairly clean. This does offer a simpler solution than trying to add a separate voltage regulator.

Development Support

As I mentioned, Microchip's PICkit 2 programmer supports these parts. So does the latest version of my PICBasic Pro compiler–except for the PIC12HV615, which has just recently been released to the public. This is the 8-pin version of the PIC16HV616, so it's got its own set of great applications. Can you imagine that part in a small surface-mount SOIC package without the need for a regulator? I can already think of several applications where that could be squeezed into a tight sensor package. It could use an A/D port to read the sensor element and then send the value back via a one-wire serial connection. This could reduce the connections in a common 12V setup to Power, Ground and Serial Signal, and not take up much space at all. Interesting, isn't it?

Now, build up your own development board with a 20-pin socket and you have a setup complete for 20-, 14- and 8-pin development. Better yet, check out the DM164120-1 development board package at www.microchipDIRECT.com. It's shown in Figure 5 and is really a great deal. You get one populated board and two blank boards for $23.99. That's about $8 per board, and it has the serial programming port already wired in.

Figure 5: Microchip's DM164120-1 development board

Another great board is the PICPROTO20 from microEngineering Labs (melabs.com), shown in Figure 6. This board has all the circuitry setup for a voltage regulator, external crystal and MCLR pull-up resistor. It even includes a header that matches their programmers for In-Circuit Serial Programming™ (ICSP™). They retail for $12.95 each, so it's not a bad way to develop with this family of PIC MCUs.

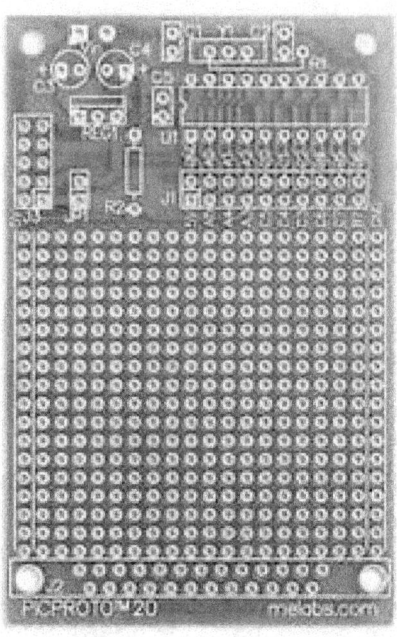

Figure 6: The PICPROTO20 from microEngineering Labs

Conclusion

I promise more projects in the near future, and some of them will include these smaller PIC MCUs. I just wanted to let you know what's out there, as you gather the pieces you need for your next PIC MCU project. I just cannot believe how easy it is getting to create unique items with a PIC MCU. I'm constantly thinking of new applications and projects, but keep running out of time to get them into a decent format for these columns. This is why I can honestly say more projects are coming, as they are sitting on my bench half complete. If you have a particular project idea you want me to consider, shoot me an email at chuck@elproducts.com, and please stop by my Web site at www.elproducts.com.

I am working on improving it with more information for visitors. I'm finding all kinds of useful links as I help Microchip customers solve their design challenges. Some of the links are just hard to find on the Microchip Web site. Hopefully, I can put links to them on my Web site so you have a one-stop place to easily find what you need.

Until next month, keep on having fun with those PIC MCUs.

June 2007 - C Language Introduction

In my April Nuts & Volts column I did a speed-test comparison of various PIC®
microcontrollers (MCUs) that were programmable in Basic, and then compared a
Basic and C compiler. I made a mistake that affected the results. I claimed that
microEngineering Labs' PICBASIC PRO compiler outperformed HI-TECH's
PICC-Lite compiler for speed, and I was wrong. PICC-Lite compiled code runs
faster. I'll explain how I goofed a little later.

The response to the speed-test article was amazing. I got more email response from
that article than any article prior. Several knowledgeable C language people
pointed out my mistake. I also had a few people point out that Revolution
Education's PICAXE programming system could be "overclocked", meaning it
could be run with an 8 MHz or 16 MHz resonator in place of the 4 MHz resonator
I used in the speed test. At 16 MHz, the PICAXE would run almost as fast as Basic
Micro's BasicATOM.

This was great feedback, since it indicates that readers of this column include
programmers from various experience levels and are reading the details. In other
words, they aren't falling asleep reading my stuff. Now let me explain the error.

Code Error Explained

If you missed the April column speed test, I just did a simple For-Loop that
continued for 255 loops, and within the loop I processed a simple math equation to
increment a variable. I toggled a pin outside the For-Loop so I could measure the
processing time on an oscilloscope, as shown in Figure 1. In the article, I made one
mistake that several C programmers pointed out. In the section where I compared
PICBASIC PRO to the PICC-Lite compiler, I declared two "int" variables in the C
compiler version of the speed-test code.

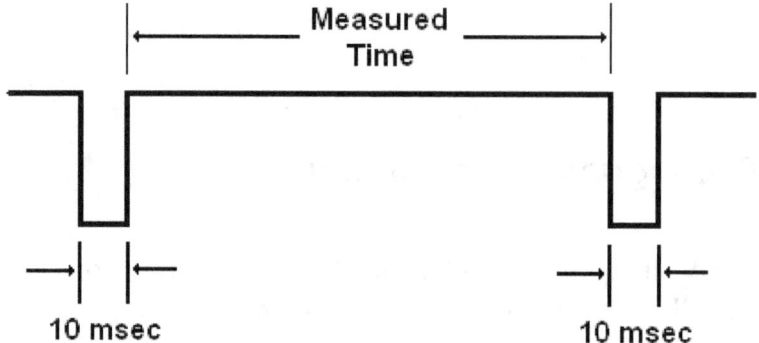

This was a huge mistake in a speed test program, since "int" creates a 16-bit variable in the C program—but in all the Basic versions of the code I declared byte variables. This greatly affected the speed results. You see, the PIC MCUs I used in the article are 8-bit microcontrollers, meaning they have an 8-bit wide data bus. To process a 16-bit variable, the program has to perform the math function in two steps—low byte then high byte. This takes twice as long. Therefore, it incorrectly showed the C compiler slower than PICBASIC PRO.

To add to the confusion, I didn't even realize until much later that I also sent the final version of the article with a typo in the C code embedded in the text. The C code in the published column included the For-Loop variable changing from 0 to less than 256; when in reality I ran it 0 to less than 255 because I knew a byte variable would always be less than 256. Changing from 256 to 255 should have tipped me off to the "int" variable error, but I completely missed it. You see, a byte variable will roll over to 0 after reaching 255—since the 256 decimal is 100000000 binary. As this shows, it takes 9 bits to store the 256 decimal and the lower 8 bits (or byte) are zeros. So, I was thinking byte variable in my corrected code, but left the 16-bit "int" variable declaration in the code. If I wasn't focusing on a speed test, it wouldn't have been an issue.

I even admitted I was shocked by the results of the PICBASIC PRO compiler being faster than C. I tested it on another processor, but got the same results since I made the same mistake. If an operating-system supplier finds a bug, then they release an "upgrade". This article is my upgrade. I'm correcting my mistake and clearly pointing out that the speed of the PICBASIC PRO was fast, but not as fast as the corrected version of the C code compiled by the PICC-Lite compiler.

After a very nicely written email from Langue Rodriguez pointed out my mistake, I went back and declared the variable in the C program as an unsigned "char" variable (which made it an 8-bit variable) and re-ran the speed test. The corrected code is shown in listing 1. This re-test showed the PICC-Lite C program was faster than PICBASIC PRO by 460 microseconds, with the setup running at 4 MHz. The PICBASIC PRO program took 1.76 milliseconds and the corrected HI-TECH PICC-Lite version only took 1.3 milliseconds.

Listing 1

```
#include <pic.h>            // Include HITECH CC header file

void Pause( unsigned int usvalue ); //Establish pause routine function
void msecbase( void );          //Establish millisecond base function

main()
{
PORTB = 0;                   //Clear PortC port
TRISB = 0;                   //All PortC I/O outputs


while(1==1)                  //loop forever
{
    unsigned int z, y;
    RB0 = 1;                 // Turn on RC0 LED
    for(z=0; z<255; z=z+1)
    {
    y=y+1;
    }
  RB0 = 0;                      // Turn off RC0 LED
    Pause(10);                  // Pause 10 msec


}       //End while
}       //end main

//********************************************************
//pause - multiple millisecond delay routine
//********************************************************

void Pause( unsigned int usvalue )
{
    unsigned char x;

for (x=0; x<=usvalue; x++)    //Loop through a delay equal to usvalue
            {                 // in milliseconds.
            msecbase();       //Jump to millisec delay routine
            }
}

//********************************************************
//msecbase - 1 msec pause routine
//********************************************************

void msecbase(void)
```

```
{
    OPTION = 0b00000001;       //Set prescaler to TMR0 1:4
    TMR0 = 0xd;                //Preset TMR0 to overflow on 250 counts
    while(!T0IF);              //Stay until TMR0 overflow flag equals 1
    T0IF = 0;                  //Clear the TMR0 overflow flag
}
```

This was a good lesson for me to have extra eyes review my code before releasing articles in the future, and I promise to do that. It also taught me something I never really thought of prior to this. Declaring the correct size variables can have a huge impact on the processing time of your code.

I remember many times when I wanted to create a variable that would contain values only slightly larger than 255. I would declare a 16-bit value or a word variable in PICBASIC PRO. Now I see how this cost me significant processing time. To understand exactly how much it cost me, I re-ran the PICBASIC PRO code with a 16 bit "word" variable declared. It was a lot slower than the C program with the "int" variable. PICBASIC PRO took 2.6 milliseconds with a 20 MHz resonator and 13.1 milliseconds with a 4 MHz resonator. The PICC-Lite compiler took 720 microseconds at 20 MHz and 3.6msec at 4 MHz. It's very clear by looking at this, the PICC-Lite compiler handles 16-bit math much more efficiently than the PICBASIC PRO.

All this speed comparing between these two fine compilers is relative, though, because—despite this 16 bit variable slow down—PICBASIC PRO with a 16-bit variable declared and running at 4 MHz was still faster than the PICBASIC standard version—even with the variable still set as a byte (reference table 1). I'm often asked, "why should someone spend a couple of hundred dollars for the PICBASIC PRO compiler when they can program with a cheaper option?" This is clearly one of the reasons.

In many projects, high speeds won't matter. But in cases where it does, it's good to know the limits—especially if your program has lots of math equations. Those applications may be the time where you use a byte variable for the low byte, and then have a second byte variable for the high byte, followed by a check to see whether the low byte overflowed before updating the high byte. This would save processing time, since you are working with byte-sized variables.

Other Compiler Choices
Another batch of comments came from people curious about other compilers, such as Crownhill Associates' Proton Basic compiler. I've tried it in the past and

remember it using a slightly different setup than Parallax's BASIC Stamp PBASIC style, which the test program was based on. I decided to try Proton on this speed-test code. I downloaded the sample version of Proton to see if it supported the same chips I used. The sample version doesn't support the PIC16F876A or PIC16F877A, but did support the PIC16F877. I dug around and found one of those older chips, so I was able to run it. I was pleasantly surprised to find the same speedtest.bas program compiled in Proton, but I got a few warnings. Proton prefers you use the DIM instead of VAR directive for creating variables, and it uses a different command than PAUSE for delays. The compiler did compile it, though, and just gave warnings. The warnings even stated the commands were recognized for backward compatibility, which is a nice feature. I only ran the test at 20 MHz on the PIC16F877. As it turned out my test showed Proton ran exactly equal to PICBASIC PRO, completing the loop in 350 microseconds @ 20 MHz.

Langue Rodriguez sent me an email that he had tested a corrected version of the C file on the CCS C compiler, and included the results in his email. He reported a result of 1.794 msec at 4 MHz with the variable corrected to 8 bits (int8). According to this result, CCS is slower than PICBASIC PRO's 1.76 msec in 8-bit mode, but I really wanted to confirm this so I ran it myself. I measured 1.792 msec when I tested Langue's code myself with the CCS PCM version and a PIC16F877A at 4 MHz. To get this accuracy, I used a much finer resolution on the scope. I went back and re-measured the PicBasic Pro code at this scope resolution and found it to be 1.790 msec. Therefore I concluded that CCS and PicBasic Pro ran the same speed and my previous measurement had some resolution error. I ran the PICC-lite at this finer resolution but got the same 1.300 msec as before. Langue's CCS code is below if you want to test it yourself.

Listing 2

```
******************************************************

c-file follows:

// Basic program to compare speed and code size
#include <16F876A.H>
#device *=16
#fuses XT, NOWDT, NOPROTECT, BROWNOUT, PUT, NOLVP
#use delay(clock=4000000)

#define RB0 PIN_B0

//======================================
void main()
{
  int8 z, y;
```

```
  while (true)
  {
   output_high(RB0);
   for (z=0;z<255;z++)
   {
    y++;
   }
   output_low(RB0);
   delay_ms(10);
  }
}
```

* *

C Introduced

This seems like a good point to introduce some simple C coding to those who haven't worked with this language. The C language at first seems a lot different than Basic, but in reality it's more format that anything. C code for PIC MCUs is different than trying to write C code for a PC application. In this respect, having the PIC MCU experience from programming in any other language is a bonus, because you understand the internal setup. For example, let's take the very simple program in Listing 3 written for the Hi-Tech PICC-Lite compiler. (It just lights an LED on PORTB bit 0.) I use it because it shows the basic structure of C coding. The top of the file has the line #INCLUDE <pic.h>. This puts all the necessary PICC-Lite compiler setup information in the program, and each C compiler will have its own version of this included file.

The next line establishes the configuration settings. I wrote this simple program in Microchip's MPLAB® Integrated Development Environment (IDE) and set it up as a project. In the project setup, I selected the PIC16F876A. The #if to #endif block of code tests for this processor selection, and if any of those listed are selected in the MPLAB IDE, the compiler will include the configuration fuses to run in the external crystal mode (XT) with watchdog timer (WDTDIS), brownout reset (BORDIS) and low voltage programming (LVPDIS) disabled. This will get transferred to the programmer automatically, through the final compiled and assembled .hex file.

The main body of the code is next. All C programs start with the main label followed by parenthesis. Below "main" is always a set of braces "{}". All code in a C file is contained within the main braces. The first operation is to setup the PORTB and TRISB registers. We clear them both to make all of the PortB outputs and cleared. This doesn't look much different than a Basic program.

64

The program then forms a continuous loop with the While statement. The formula in the parenthesis is tested and, as long as it is true (or results in a logical 1), then the commands within the While statement's set of braces are executed over and over. This is also not much different than the While command in many Basic compilers, except the commands are not contained within braces. In this case, the formula is 1==1, which is different than using a single equal sign. The double equal tests whether the right side is equal to left side, and if they are equal then it is considered a true statement—so a "1" is returned. Since one will always equal one, everything in the While loop will occur over and over again. The command within the While braces sets RB0 or PortB bit 0 to a one, which turns on the LED.

The program is completed by putting the closing braces for both the Main and While loops. It closes with an end command.

One thing to note with the structure of C is that all command lines end with a semicolon. This is one of the biggest errors for beginners, as it is not required in assembly or Basic. Also, comment lines start with a double backslash—not a colon or semi-colon like assembly or Basic. If you can get past the braces and colons, the program doesn't look much different than a PICBASIC program. You can see some of the differences between the Hi-Tech PICC-Lite and the CCS C compilers by comparing Listing 3 to Listing 2. I mainly wanted to introduce the basics of C code, and plan to help you learn more as this column progresses along.

Listing 3

```
#include <pic.h>          // Include HITECH CC header file

#if defined(_16F873A)   || defined(_16F874A)     ||\
    defined(_16F876A)   || defined(_16F877A)
__CONFIG (XT & WDTDIS & BORDIS & LVPDIS);
//XT Crystal Mode, Watchdog off, BrownOut Disabled
#endif

main()
{
PORTB = 0;                   //Clear PortC port
TRISB = 0;                   //All PortC I/O outputs


while(1==1)                  //loop forever
{
    RB0 = 1;                 // Turn on RC0 LED

}      //End while
}      //end main

end;
```

Conclusion

I still feel that programming PIC MCUs in Basic is the best option for the beginner, but if you want to do any programming outside of the hobbyist arena you should learn at least some C. I started with assembly and have used Basic for years, so those are like second nature to me. Adding C to my software skills made programming PIC MCUs even more fun, and I hope to pass on what I know to those that want to learn C as well.

I've watched and helped many complete beginners get started programming by using the BASICATOM and PICBASIC PRO. I also know others who learned with the various other options. There is no single correct choice. If Proton or PICC-Lite or PICAXE or Basic Stamp or any other choice gets you started programming PIC MCUs, then go for it. When I started, the choices for the hobbyist or beginner were limited and finding help was next to impossible. I often wished that I had someone teaching me or had books to read when I got started, but I couldn't find many. Those books I did find were written for the professional and were often way over my head. I had to find all this out on my own. I hope you keep reading, and I'll be more careful about the errors. Send me emails with your feedback, because as you can see I listen. Send to chuck@elproducts.com and stop by my website anytime at www.elproducts.com.

July 2007 - Hot Wheels Drag Race Finish Gate

On a recent trip to the local Kmart I saw that Mattel had released a series of classic Hot Wheels sets similar to the ones you could buy when I was a kid. I bought one of the two lane drag race types with the mechanical flag that flipped one way or the other to indicate which car got to the finish line first. It was the classic drag race toy from my youth. My daughter was fascinated with it but had a hard time resetting the flag. It would also stick sometimes and not indicate any winner. I decided this was a good time to replace mechanical with electronic and I chose the Atom as my solution.

An electronic drag race finish gate will be the project this month along with a demonstration of how to use a very simple setup to program an Atom chip.

Atom Chip

For those not familiar with the Atom family of parts, they include three different chips, the Atom 28A, Atom 28B and Atom 40. They are based on the PIC16F876A (Atom 28A,B) and PIC16F877A (Atom 40). They also are available in module form similar to the Basic Stamp with programming interface circuitry and voltage regulator built into the module circuit board. These modules are all surface mount so they can fit in a small package but I prefer to work with leaded components so I can build it on a breadboard and easily replace parts if I fry something. It's also cheaper to use the chips rather than pay for the regulator and programming interface on every project.

My ultimate OEM module and BasicBoard module use the leaded versions of the chip but that again may be more than the beginner is willing to invest to get started programming. The truth is the modules just make it easier to use the Atom chips. All you really need to use the Atom chips are a 5v source and an RS232 PC serial connection to program the Atom chip.

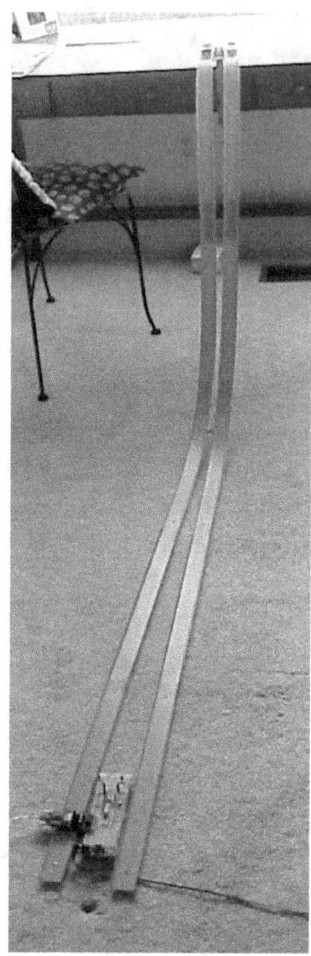

Figure 1

For the drag race finish gate I'm going to use an Atom 28B chip which is the same one I use in the Ultimate OEM module. The finished setup is shown in Figure 1. The schematic for the setup is shown in Figure 2. The Atom28B and Atom 40 both use one pin to communicate to the PC. That single pin does both the Tx and Rx function but requires a diode between them to make it work. A 10k pull-up resistor is also recommended on the single line after the diode. Basic Micro did this so it would only steal one I/O pin for programming and communicating back to the PC when running in Debug mode. You see the Atom will allow you to step through your code command by command so you can watch the variables change along with the internal registers within the chip. This is a great feature when you need to figure out why your code isn't doing what you expect.

The Atom 28A chip was the original Atom chip and it uses two I/O pins for communicating and programming. This doesn't require a diode but two 10k pull-up resistors are recommended. Therefore you lose an I/O port and still add two external components.

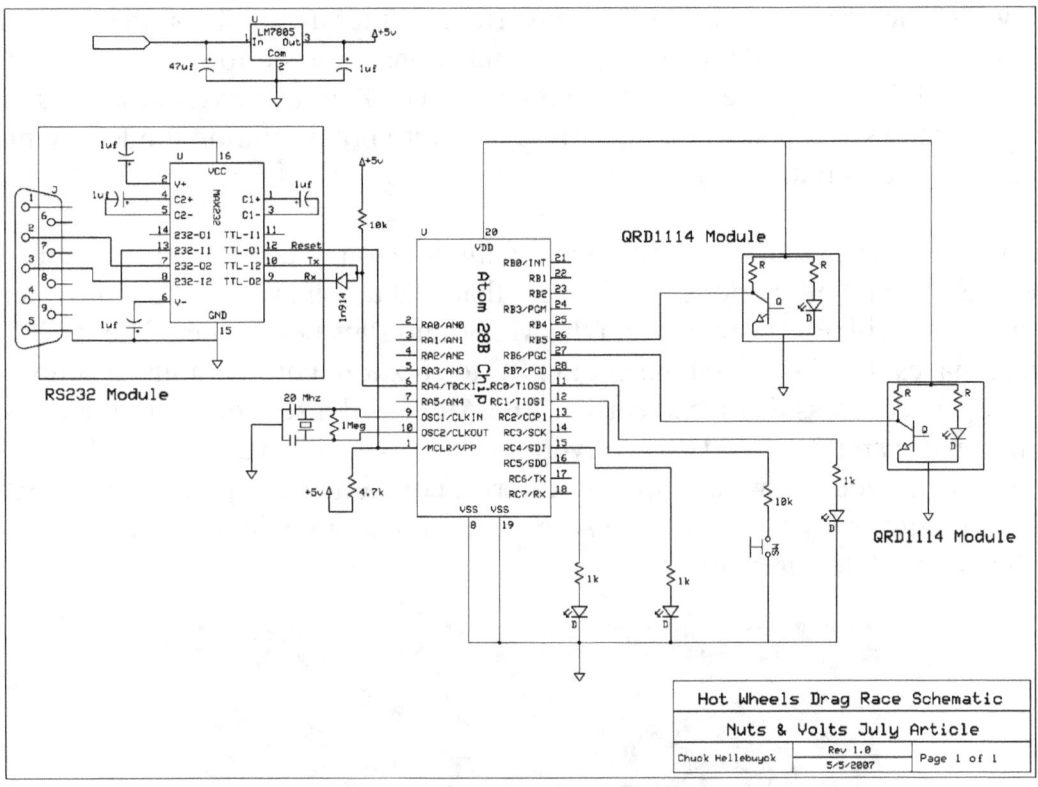

Figure 2

Atom Programming Setup

The Atom chip has a custom bootloader pre-programmed in that will download the program from the PC to its own internal program memory. It does this thru the the Atom compiler which actually produces a hidden .hex binary file similar to any compiler such as PicBasic Pro or even Basic Micro's own MBasic compiler. In fact, the Atom software is a custom version of their MBasic Pro compiler limited to the Atom chips. After writing your code you simply click the "Program" button in the Atom IDE screen and the software will compile the Basic Language code, produce a .hex file with the binary 1's and 0's and send it to the Atom chip thru the PC's RS232 port and into the Atom chip via the bootloader. The Atom chip will receive the .hex file and program its own program memory. The program will start running as soon as the download is complete.

You can't just connect the Atom chip to the PC serial port though. The communication between the Atom chip and the PC requires a level shifter circuit to convert the +12v, -12v RS232 signals into the 0v, 5v signals the Atom chip can work with. There are various RS232 converter modules available on the internet but most don't include all the necessary connections required for the Atom. Most of these modules only have the Tx and Rx pins. The Atom software also uses the DTR pin of the serial port to put the Atom chip into programming mode/run mode. Just having a reset button won't work.

I designed my RS232 Breadboard interface module to include the reset feature since I also use it with various bootloaders that will automatically reset the micro. I use this setup with PicBasic Pro and the MCStudio Plus bootloader. The RS232 module makes it real easy to connect the PC to an Atom chip with just a couple simple connections as described earlier. The Atom will also work with the USB port if you use an RS232 to USB converter cable. The schematic shows the connections so you can wire it up with discrete components in place of my RS232 module. In Figure 3 you can see the Atom 28B chip along with the RS232 breadboard module wired up.

Figure 3

Sensors

I wanted to make my system non-contact so it would not interfere with the Hot Wheels cars racing down the track. One of the changes they made to the classic track was to add holes at both ends of the flat track. These are there for custom track connectors but when the track ended up on flat ground they weren't used. I decided to find a sensor to fit in that hole. At first I thought about using a CDS cell but that would require a separate light source to shine from above. Then I remembered my servo sensor design I use to monitor rotation of a robotic wheel (figure 4). It uses a QRD1114 reflective sensor.

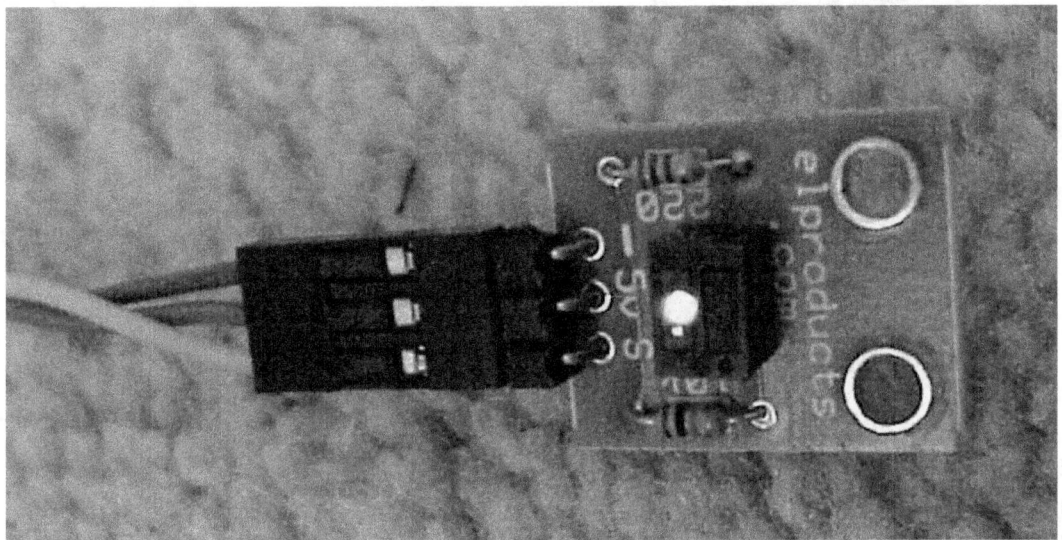

Figure 4

These sensors have an LED and light detector combined in one package. The LED sends a beam of light out and the sensor picks up the reflection if something other than a black colored object is in front of it. I developed a little module that simply powers the LED and adds a pull-up resistor to the output. When the sensor sees black or no reflection the output is high. When the sensor sees white or a reflection the output goes low. I decided to make the hole in the track a little larger so one of these sensors could be placed in each track (figure 5). A close up of the finished setup is shown in figure 6.

Figure 5

After a quick test of running a couple cars down the track proved the sensors did indeed sense when the Hot Wheels cars passed over I now had to connect them to the Atom chip and write the software to determine which one saw the car in its lane first. I decided to just have the Atom light an LED on the left or right to indicate which side won the drag race. The hardest part in my mind was reading the sensors at the same time, deciding if there was a car sensed and then which one won and do it all within a very short period of time that the cars where passing over. As it turned out the Atom handled the challenge just fine. I also wrote the software to capture the state of the sensors to give the program more time to sort out the winner.

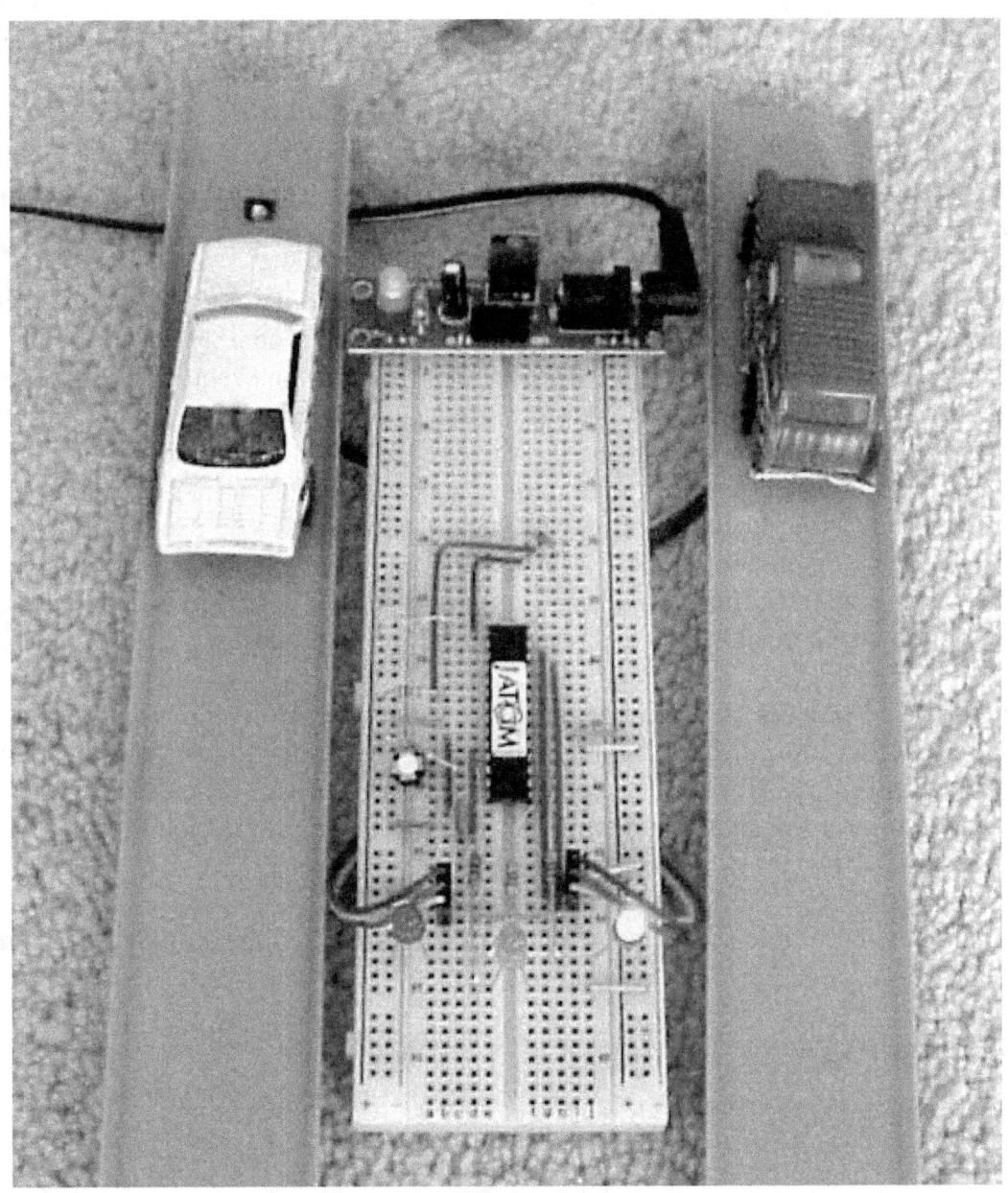

Figure 6

Hardware

The hardware is a combination of what I already described plus a little bit more. The Atom chip programming setup as described already plus the QRD1114 sensors connected to PortB pins P5 and P6 make up most of the circuit. The reset of Port B is left unconnected but could be added to additional sensors for more lanes if needed. In the software I enable the internal pull-up resistors on Port B so all the Port B pins will see a high when no car is sensed. If one of the sensors sees a car

73

then either the P5 or P6 input will go low. This makes the software easier since all the program needs to look for is PortB to drop below FF hex (or 11111111 in binary) to then jump to the section of code that determines which car won. I'll cover more on that later in the software description. The LED's that indicate which lane won are just lit through a 1k resistor. The extra LED is used to indicate if the setup is armed and ready to look for a winner.

I used one of my breadboard 5v regulators to power the rails of the breadboard but any 5 volt source will work. The Atom needs a 20 Mhz resonator and the typical MCLR pull-up resistor which is 4.7k value. The transmit line requires a 10k pull-up and the diode is needed for the Atom 28B chip as described earlier. The reset line from the RS232 module connects to the MCLR pin. So if you add it up, the Atom chip only needs two additional connections; Tx and Reset, to program the chip since the rest will be required for the circuit to work. If you use the 28A chip you'll need three connections (Rx, Tx and Reset). The advantage to this setup is the RS232 module can move from board to board and all you need is a different Atom chip on each design. This is cheaper than buying modules.

The QRD1114 sensors are connected through servo extension cables and then through a header into the breadboard. The re-arm switch is tied to Port C1 which requires a pull-up resistor so I used a 10k. This way the switch is pulled high until it's time to re-arm the setup. The armed and ready LED is connected to Port C0 through a 1k resistor.

That's it. A very simple circuit because the complexity is in the software. Or is it?

Software Listing

```
setpullups pu_on            'PortB Pullups On
stat var byte          'Create variable Status
trisb = %11111111      'RB6 and RB7 input, Rest of PORTB Outputs

main
        high p8                         ' LED On indicating Armed and ready
        stat = portb                    ' The value of PortB is captured
        if stat = $FF then main         'Test for Car, loop back if not sensed
        if stat = 223 then              'Test if Right Lane won
                high 12                 ' Right LED on
                low 13                  ' Left LED off
                goto wait               ' Jump to reset loop

        elseif stat = 191               'Test if Left Lane won
```

```
            high 13              ' Left LED on
            low 12               ' Right LED off
            goto wait            ' Jump to reset loop

     else
            high 12              ' Tie
            high 13              ' Light both LEDs
            goto wait            ' Test again

     endif

wait
     low p8                      ' Armed LED off
     if in9 = 1 then wait        ' Wait for rearmed switch press
     low 12                      ' Reset LEDs
     low 13
goto main
```

How It Works

The software starts off by enabling the internal pull-ups on Port B. This is the only port on the Atom 28B chip (PIC16F876A) that offers this feature. The ports also have to be in input mode for the setting to take effect. That is the next port control instruction where the TRIS register for Port B is set to all 1's making all the Port B pins inputs. In between these two commands the "stat" variable is created.

```
setpullups pu_on            'PortB Pullups On
stat var byte               'Create variable status
trisb = %11111111           'RB6 and RB7 input, Rest of PORTB Outputs
```

The main loop is the next section. The armed and ready LED connected to the C0 pin is set high. The Atom software has predefined nicknames set for the port pins. Port B is P0 thru P7, Port C is P8 thru P15.

```
main
     high p8                     ' LED On indicating Armed and ready
```

The value of Port B is captured and stored in the variable "stat". The signals at Port B will change quickly so capturing and storing the value in a variable allows the rest of the program more time to read and respond to the results.

```
     stat = portb                ' The value of PortB is captured
```

All of Port B is set to inputs with the pull-up resistors turned on so Port B should look like all 1's if nothing is connected. The sensor modules pull the line low when a car is sensed so when sitting idle those inputs look like 1's also. Therefore, this command just looks for a value less that $FF hex (or %11111111 binary) to determine if a car has passed by. If nothing is sensed it loops back to the "main" label and then checks again.

```
if stat = $FF then main     'Test for Car, loop back if not sensed
```

If a value less that $FF was detected, the program checks which bit changed indicating which lane won. If the value is 223 decimal or binary %11011111 then the right lane sensor was tripped. The right LED is turned on and the left is turned off. Then the program jumps to the wait label.

```
if stat = 223 then          'Test if Right Lane won
        high 12             ' Right LED on
        low 13          ' Left LED off
        goto wait           ' Jump to reset loop
```

If the right lane was not tripped then the program tests the left lane bit. It would equal 191 decimal or %10111111 binary if the left lane sensor was tripped. If this happened then the left LED is turned on and the right one turned off.

```
elseif stat = 191           'Test if Left Lane won
        high 13             ' Left LED on
        low 12          ' Right LED off
        goto wait           ' Jump to reset loop
```

The final step of the If-Then-Else command is the "else" section. If both bits were tripped or a strange result was seen then both LEDs light to indicate a tie or something wrong but either way the race must be run again to determine a winner.

```
else
        high 12             ' Tie
        high 13             ' Light both LEDs
        goto wait           ' Test again

endif
```

The last section in the program starts at the "wait" label. The first thing that happens is the armed LED is cleared and then the program goes into a loop. It tests if Port C1, or P9 in Atom nicknames, is low. A low indicates the re-arm switch was pressed. If the switch isn't pressed the port stays high because of the 10k pull-up resistor shown in the schematic. When the switch is pressed, the program clears the winning lane LED and the programs jumps back to the "main" label to wait for the next race.

```
wait
        low p8              ' Armed LED off
        if in9 = 1 then wait ' Wait for rearmed switch press
        low 12              ' Reset LEDs
        low 13
goto main
```

Next Steps

There are so many ways this program can be expanded. More lanes could easily be added which would be tough for a mechanical system. I also want to add timing so each lanes time is displayed on an LCD screen. This will take a little more strategy in the way the program responds to a sensor signal though. A timer value has to be captured for each lane. Another point is the sensors that I used rely on the bottom of the car to be reflective. As long as the bottom isn't black it seemed to work fine. But one of the cars tested was all black on the bottom. A sensor above the cars would have probably worked on that car but I had enough cars to test so it wasn't a big deal to throw that car out of the pack for now. Possibly a better sensing method would be worth pursuing.

I also would like to add an automatic launch system so a button can be pressed and then some kind of motor or solenoid releases the cars automatically. This will probably be part of another article. Feel free to try these ideas out yourself and email me your results. I can always be reached at chuck@elproducts.com.

Conclusion

Projects like this are real important to me even though it may seem simple or rather dumb to someone looking for more complex applications. I feel projects like this are important because anyone can understand the benefits of adding electronics over the mechanical method in a simple toy setup like this. It may not be a cheaper solution but to someone who is trying to learn programming, it's projects like this that make a connection. Show a complex project to a kid and he may be amazed at what you did but deep down not have a clue what you are talking about. But show a kid something like this and they fully understand what you did. That could be the connection that gets them interested in electronics or technology in general.

I'm not claiming I made that connection happen but maybe a few readers will try this and some kids will see it and down the road want to learn programming as well. That is worth far more to me than creating some complex programming algorithm. Take it further by laying out a board and turn it into a kit. This could be a great project for a Boy Scout troop or science fair demonstration.

I also wanted to mention that in August Microchip is having it's annual Master's conference. It's a 3-5 day affair with loads of classes that teach everything from Ethernet to C Programming to just learning about the various PICmicro MCUs. I'll be there as part of the Microchip team. If you attend, look me up. It's well worth the cost if you need development tools. Attendees can purchase development tools for half price.

As always, give me your feedback at chuck@elproducts.com or visit my website www.elproducts.com. I have the Atom chips, RS232 modules, 5v regulator modules and breadboards in stock if you want to reproduce the setup. The servo sensors are an item I've had on the shelf for while but haven't added them to the website yet. Guess I need to get to that. See you next month.

August 2007 - Basic ATOM In-Circuit Debugger

If you've thought about getting started with PIC® microcontrollers (MCUs), you've probably seen the variety of available development tools, such as programmers, debuggers and the Integrated Development Environments (IDEs). To a beginner this can be very confusing, because there are Microchip's own tools and also many from third party suppliers—and the prices range from a few dollars to $1000's of dollars. I get emails from beginners who have seen this and wonder what they should buy. They want to get started, but don't want to spend money on a tool and then find out they didn't need it or later learn that it was intended for the advanced user. One of the reasons that I like to recommend Basic Micro's BasicATOM chips for the beginner is because of the great In-Circuit Debugger (ICD) built into the BasicATOM software.

We all can understand what a programmer does, and I've talked about various options in previous articles. The IDEs are typically free to download and each compiler has its own version. Microchip has an IDE that works with all PIC MCUs and all the Microchip development tools, called the MPLAB® IDE, and it's free to download.

Debuggers are a bit harder to understand for the beginner and, often times, are some of the most expensive tools available. They are priceless, though, if you are trying to understand why your code doesn't work properly. They offer the ability to step through your code, command by command, or set breakpoints. Breakpoints are like stop signs to software. They will halt a program at the place where the breakpoint is set. Typically, you would set a breakpoint at the section of software where you suspect an error. Once it stops there, you can single step through the commands to see if it's doing what you expect. Debuggers also will display the value of the variables in your software and special function registers inside the PIC MCU.

Using a simple BasicATOM DEBUG command will allow you to send variable data out of the serial connection to a PC screen but, being able to see deeper inside the chip and step through your code slowly as you watch the hardware react, really helps get to the root cause of your code problems. It also gets a bit tiresome having to inject DEBUG commands throughout your program, as you try to figure out what's going wrong. The BasicATOM chips and modules offer a built-in ICD feature that displays all the inner registers and variables; plus, it includes the single stepping, breakpoints and other typical debugger features. The best part is that the BasicATOM debugger is included for free. All you need to do is download the BasicATOM programming software (IDE) from the BasicMicro.com Web site, and connect your PC to a BasicATOM chip or module, and you're ready to go. This article will cover the debugger in a little more detail, so you can see how great this feature is.

In-Circuit Debugger (ICD)
The BasicATOM ICD is like a special tool that makes your software run in slow motion to make it easier to determine where an error might have occurred. Once you use it, you won't ever want to develop without a debugger again.

The BasicATOM ICD is completely controlled by software. No additional hardware is required. When you have completed a program, you would normally press the "Program" button to compile and download your program into the BasicATOM chip/module. To use the ICD, you simply press the "Debug" button, instead. The difference between the "Program" button and the "Debug" button is hidden. They both compile and then program the BasicATOM chip/module but the "Debug" adds another hidden step—it adds a block of code to your program that is used by the ICD.

When your program is running, the added ICD block of code sends variable values, internal register values and other details to the PC through the programming cable. The ICD, which is built into the programming screen IDE, will display that data in the way you choose—by clicking on the different setup buttons in the ICD debug tool bar. The debugger controls allow you to run the program continuously or in animate mode, which automatically steps command by command in slow motion through the program. You can also manually step through your code command by command, by using the PC mouse to advance the program via the various single-step buttons. This gives you total control of how the program advances. Figure 1 shows the BasicATOM IDE with the debugger enabled.

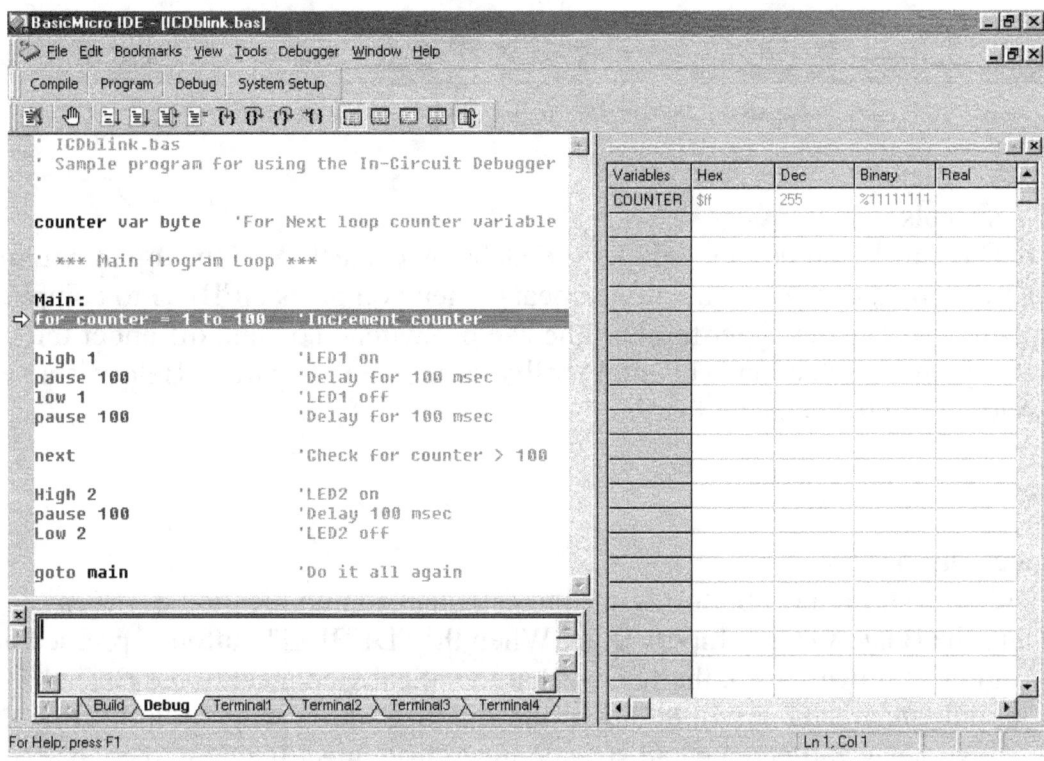

Figure 1: The BasicATOM With the ICD Enabled

As mentioned, the ICD allows you to view the state of each variable and each output state, and even allows you to monitor the inner workings of the BasicATOM chip/module (it helps to have knowledge of the PIC16F876A or PIC16F877A microcontroller, from which the BasicATOM chip/modules BasicATOM Chip is built). The ICD screen shown in Figure 1 is a program running in DEBUG mode.

Important Note
When the ICD is running in the BasicATOM chip/module, the running program can have an added delay, from 0.5 milliseconds to 500 milliseconds, depending on the action requested in the ICD. You must take this into account when running time-critical code. Each command will run in full runtime mode (SERIN and SEROUT will function normally), but added time will appear between commands. Also, the programming cable must be connected to the PC, or the Debugger will not operate and neither will the BasicATOM chip/module.

Figure 2: The BasicATOM ICD Toolbar

ICD Controls

The ICD controls can be found under the Debugger selection in the top menu line, or via the debugger toolbar line that appears when you press DEBUG to compile your program. The debugger toolbar line can be switched on and off under the View main menu selection. The total toolbar is shown in Figure 2. Below is a summary of the ICD control features.

Connect/Disconnect

The Connect/Disconnect button is used to establish communication between the ICD and the BasicATOM chip/module. When the "DEBUG" button is pressed and the program is downloaded, the ICD will automatically connect to the BasicATOM chip/module. A green bar will highlight the first line of the program indicating that the ICD is successfully connected. The connect/disconnect ICON will change to Disconnect, so you can disconnect the ICD from the BasicATOM chip/module at any time by just clicking on this icon.

Toggle Breakpoint

The Toggle Breakpoint button allows you to turn a breakpoint on or off at any point in the program. A breakpoint is a highlighted line that will stop execution of the program when it gets to the designated command line. This is handy if you want to see what the variables and I/O pins look like when a specific command is encountered in the program, without having to step through each command. To use it, just position the cursor on the line at which you want the program to stop. If a breakpoint is not set on that line, click on this icon or right click on your mouse and select "Toggle Breakpoint". This will highlight the designated command line in red and enable the breakpoint action. To turn it off, just click on the icon again or right click to turn it off.

Animate

This is a nice feature of the ICD. The Animate function will automatically step through your program, command by command, in slow motion. Each command being executed is highlighted in green. When the command is completed, the next line is highlighted. This allows you to watch and verify that the program is flowing where you expect it to go. If the "Auto Update" feature is selected (described below), then variables and internal information will be updated after each command. To view those values, though, it is often best to pause the program, as the animate mode can sometimes run too fast to allow you to read the data.

Run

This option allows you to run the program in the BasicATOM chip/module at full speed (minus a minor delay for the debugger block of code), without stopping to check for variables or other data. The green command-line indicator will not step through each command. It will just stay at the last line executed, before the Run icon was pressed.

Reset

Reset is used to start the program at the beginning. Any information stored in variables is not erased. This is a simple way to start at the beginning or to see how your program will react if a hardware reset were to occur.

Pause

The pause button will halt the program at the current command line. To resume execution, the RUN or ANIMATE button is pressed. The PAUSE button is handy to stop the RUN or ANIMATE mode, so variables and other data can be viewed.

Step Into

This is the button you press to step through your program, command by command, using your PC mouse.

Step Over

This button is a special step button that allows you to jump over a part of the program, such as a gosub or for-next routine. Sometimes a gosub or for-next routine will take many clicks of the mouse to get through the routine, using Step Into. This allows you to jump over the designated routines and move on to the command lines after them.

Step Out

This is another special step button that allows you to leave a gosub routine. It's handy for looking at part of a gosub routine, and it lets you leave when you have seen enough. Clicking this will jump you to the command line after the end of the gosub routine.

Run To Cursor

Clicking on any command line in the program will produce a blinking cursor. If you then click on the "Run To Cursor" button, the program will execute in "RUN" mode until the cursor line is encountered. The program execution will stop at that command line.

Show Variables

This control button will toggle the Variables window open or closed. When it's selected, a separate window will open and the variables defined in your program will automatically be listed. The values of those variables will be displayed in HEX, Decimal and Binary formats. (Make sure auto update is selected, so these are updated after every command).

SFRs stands for Special Function Registers. These are special internal locations within the BasicATOM chip that indicate how the internal program is responding to your modifications of the internal PIC MCU registers. This is really a function for the advanced user, but can be handy for understanding how the BasicATOM Basic program controls the PIC MCU.

Show RAM

This feature shows all the Random Access Memory in the BasicATOM chip/module, not just the variables. This is also handy for the advanced user to see the inner workings of the PIC MCU.

Show Gosub Stack

This displays the Gosub Stack. The gosub stack is the list of location pointers within the Microchip PIC MCU that directs where to jump when a gosub command is encountered. By monitoring this, you can make sure that multiple gosubs are not somehow getting lost. This is really an advanced user function.

Set Auto Update

This should always be selected. It tells the ICD to update the variable, RAM, SFRs and Stack after every command is executed. You should select this when the debugger is first connected, but it can be turned on or off anytime.

ICD Example

I want to show an example of using the ICD. This is a very simple program written to flash the LEDs on my Ultimate OEM module with the BasicATOM 28B chip installed. The program will flash LED1 on the Ultimate OEM 100 times and then light LED2, before looping back to do it all again. The variable "counter" stores the number of flashes, so we can use the ICD variable window to watch the "counter" variable value change. The program is listed below.

```
' ICDblink.bas
' Sample program for using the In-Circuit Debugger
'
counter var byte      'For Next loop counter variable

' *** Main Program Loop ***

Main:
For counter = 1 to 100    'Increment counter

High 1                    'LED1 on
```

```
Pause 100          'Delay for 100 msec
Low 1              'LED1 off
Pause 100          'Delay for 100 msec

Next               'Check for counter > 100

High 2             'LED2 on
Pause 100          'Delay for 100 msec
Low 2              'LED2 off

Goto main          'Do it all again
```

Entering ICD Debug Mode

To get started, the program is typed into the editor window. When that's done, I make sure the Ultimate OEM BasicATOM module is connected to the programming cable and is powered up. Next I press the "DEBUG" button. If everything compiles and programs properly, the debug mode will appear with the first command line highlighted in green.

Step Into

I want to see the variable "counter" change, so I press the "auto update" button and also press the "show variables" button to open the variables display window. The variable "counter" should appear in the window. Next, I press the "step into" button, to advance the program. If I keep stepping through the program, LED1 will turn on and then turn off several times. After you've seen the LED flash on and off several times, look at the "counter" variable in the variable window and see if the value of the counter has changed to match the number of times the LED has flashed (it may be one larger, depending on where you stopped the program).

Animate and Breakpoint

Now, I set the cursor to the command line with "next" in it and press the "Toggle Breakpoint" button. The "next" command line should turn red, as seen in Figure 3. Once that is completed, I press the "Animate" button to make the program run. The program should step through the main loop and then stop at the "next" command line. Click on "animate" again, and the program will once again stop at the breakpoint. Watch the variable counter change with each stop at the breakpoint.

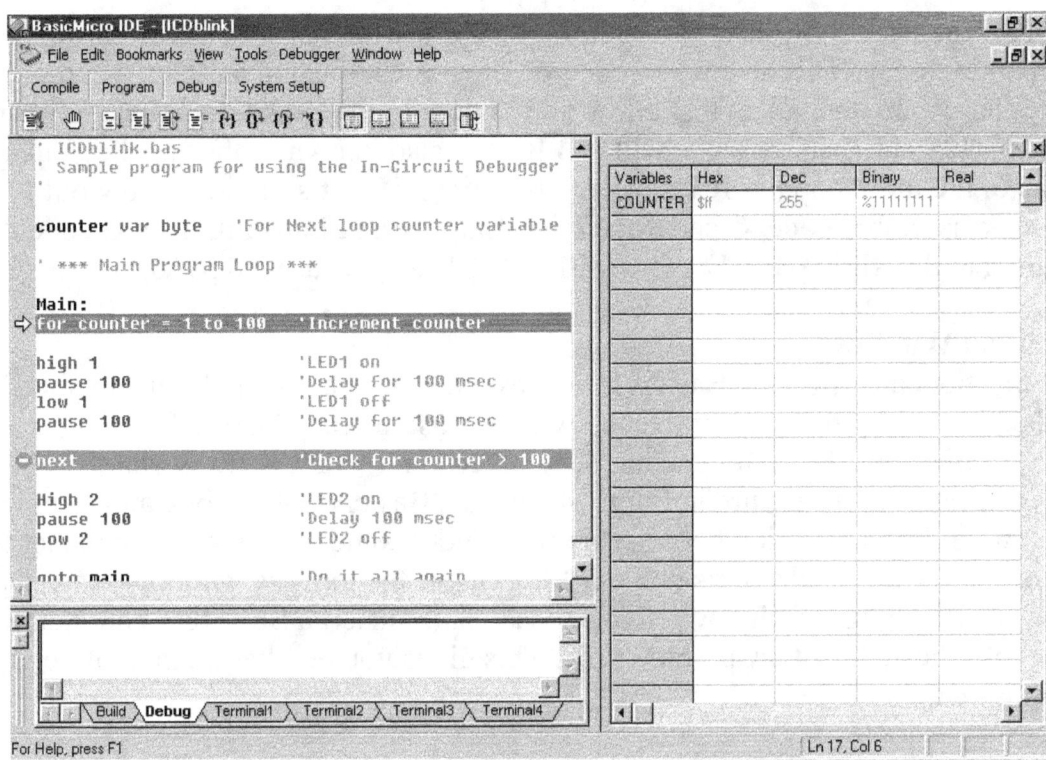

Figure 3: A "Toggle Breakpoint" Screenshot

Reset and Step Over

If I click on "Reset", the program will jump back to the beginning. Pressing "Step Over" will make the program jump past the For-Next loop and stop at the "High 2" command line. This is how you bypass a lengthy section of code to see other sections run.

ICD Summary

It's hard to present an example like this in an article, but hopefully you understood what I was trying to demonstrate. The example presented above is quite simple, but hopefully you can see how handy the ICD can be for debugging your program. As your program grows, the complexity also increases. Trying to find out why a program that runs, but isn't doing what you expect, is almost impossible without some help. The ICD is priceless for this, and yet it's included free with the BasicATOM software. Play with the debugger often to understand all its features. You can set multiple breakpoints or run to cursor multiple times, which is incredibly handy.

The ICD makes the BasicATOM chip/module one of the best beginner set-ups on the market, today. This also will teach you the basics of using an ICD. Microchip makes hardware debuggers that allow you to do all of the same functions in your PIC MCU circuit. Microchip's MPLAB ICD 2, also known as the "hockey puck", is one of the more popular ICDs offered by Microchip. It's the way you would debug an assembly code, C code or a microEngineering Labs PICBASIC coded program in the MPLAB IDE. (The MPLAB ICD 2 costs around $160.)

Terminal Window

As I mentioned earlier, the BasicATOM has a DEBUG command. This method sends data serially to a built-in Debug window also included in the free BasicATOM IDE. The IDE also has several terminal windows for serial communication. The picture in Figure 4 shows a BasicATOM debug and terminal windows. It has a selection bar to setup the window, and allows you to connect to one of the various COM ports on your PC. Any of the terminal windows can also be used as a debugger window, if you occasionally insert a SEROUT command that sends the status of a variable or pin. This gives you another method of debug, in addition to the ICD.

I use these windows often, but typically not for debugging, since the ICD works so well. Additionally, I often use these windows for two-way communication between the BasicATOM and the PC. In fact, I demonstrate this in one of the projects in my BasicATOM book. In Chapter 13 of my book, "Programming the BasicATOM Microcontroller", I use this window to communicate with the Ultimate OEM module and remotely control which I/O pin to drive high; thereby lighting one of eight connected LEDs.

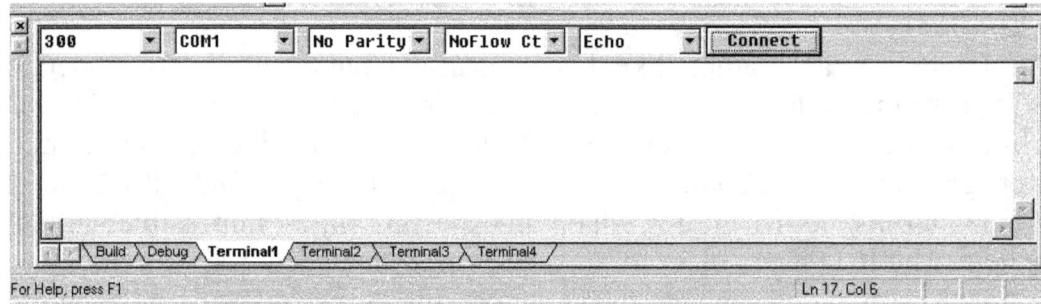

Figure 4: A BasicATOM Terminal Window

Terminal Setup

The terminal window can be setup in various formats. The most commonly modified option is the baud-rate selection, shown in the first pull-down box. Here, you match this up to the baud rate that your SEROUT command uses.

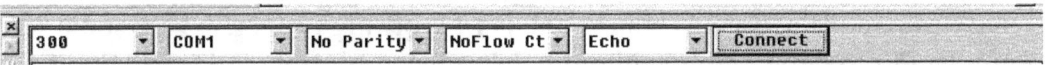

Next, you can select the proper COM port and then setup the other communication parameters, such as Parity or No Parity, flow control or no flow control. Also, you can set it to echo back and display what you type within the terminal window, after you hit the enter key.

The last item is the "Connect/Disconnect" button. When you are ready to receive or send data to a BasicATOM module, you would connect by clicking on the Connect button. You won't be able to program the BasicATOM if you use the programming port S_IN and S_OUT in your SEROUT command as the communication port, until you disconnect—since these are the same connections used to program the BasicATOM. Using those pins, though, makes it easy to use—since those pins are already connected to the RS232 level shifter chip in the Ultimate OEM module and the PC.

You can also have more than one terminal window running, by selecting them from the tabs at the bottom of the terminal window, as seen below.

The terminal window is a handy tool, and having it as part of the BasicATOM IDE is very handy. If you like to create projects that involve PC-to-microcontroller communication, you will use it often.

Conclusion

I find that the BasicATOM ICD is just a great feature, no matter how you look at it. In fact, it's one of the features that helped sell me on the BasicATOM. If you are trying to teach someone programming, being able to step them through a program while the hardware changes for each command makes it much easier to understand. The sample program I showed here flashed two LEDs on and off. Being able to watch the LED turn on and off after each HIGH or LOW command, respectively, is a great way to teach programming to the beginner.

When you get a program that acts strange—it looks like it should operate one way, but it does something different—being able to see the various variables and inner registers may tip you off as to why your code isn't working properly. I've spent many a night trying to debug code without a debugger tool. It's really a difficult task.

Big News
In addition to showing you the great debugger feature of the BasicATOM, I can finally officially announce something I commented about in a previous article. By the time this article hits the newsstand, you should see that new BasicATOM chips are available based on the new PIC16F886 and PIC16F887. These are the next-generation PIC MCUs that improve on the PIC16F876A and PIC16F877A, upon which the previous BasicATOMs were based. Basic Micro also lowered the price significantly.

One of the complaints I had about the BasicATOM was the price of the BasicATOM chips. If you wanted to create a product based on the BasicATOM, spending $20 on a chip was too expensive. As I write this the BasicATOM chips are planned to sell for around $10 and possible quantity discounts. This is outstanding news, because the BasicATOM gives you so many features—including floating point math and commands, which even PICBASIC PRO doesn't offer. Check out these new BasicATOM prices, and if you are just getting started give the BasicATOM chips another look. I honestly do not see a better deal out there for the beginner, or even an experienced user, who needs a quick way to develop a product.

As always, contact me at chuck@elproducts.com with feedback on this column. I try to read them all and answer them as fast as I can. You'll also notice my Web site shifting from selling products to just supporting my books and articles. I just could not keep up with it all, so the selling part of my Web site is going to be shifting to other sources (so I can focus on writing more books and articles). Visit anytime you want at elproducts.com, as I hope to add more beginner tips to the website.

September 2007 - Large Digits on a 4x20 LCD

Driving a Liquid Crystal Display (LCD) module has become very easy to do with the various PIC® microcontroller (MCU) options that are available. microEngineering Labs' PICBASIC PRO compiler, Basic Micro's BasicATOM chips and many other PIC MCU-based chips, modules and compilers offer a command dedicated to driving these types of displays. When I've discussed driving an LCD module from a PIC MCU in the past, I received several emails asking where to buy one. I just took it for granted that everybody knew they were practically available anywhere. The secret is in the common driver chip used in the LCD modules. LCD modules are 99% of the time controlled by a Hitachi 44780 chip that handles all of the character generation. If you find an LCD module with that chip, then you are set. You can get that type of LCD module from all of the typical suppliers, such as Digi-Key, Mouser, Jameco and others, but you can usually find a better deal by visiting individual Web sites. One of my favorites is junun.org. They sell robot kits and parts, and also have a great deal on LCDs. You can even get surplus LCDs and LCDs removed from equipment, at Marlin Jones (MPJA.com) and others, at incredibly cheap prices. I use a surplus unit in this month's project.

For this month's article, I wanted to cover a little different feature of the Hitachi LCD chip's character generator, which many may not know about. The first eight locations of character memory can be modified to make custom 5x8 pixel characters. In this month's column I will load custom 5x8 characters into those eight locations and then use those custom 5x8 characters to display large characters on a 4x20 LCD. I will drive the 4x20 LCD from a BasicATOM 28-pin chip, which is just a PIC16F876A with the custom BasicATOM self-programming bootloader installed. To make the connections simpler, I use my Ultimate OEM module with the BasicATOM 28 installed. However, wiring it directly to any PIC MCU will also work. If you want to use the PICBASIC PRO compiler or some other compiler, even the software setup is very similar. Let me show you how it's done.

Project Description

To start, let me explain how the custom characters are set up in the LCD. The LCD control chip has eight locations at the beginning of its character memory, which can be modified to make custom 5x8 characters. Once those are stored in the LCD's chip, these custom 5x8 characters can be called the in the same manner as any standard, pre-stored ASCII character. In Figure 1, a 5x8 character bit map is shown with the first custom character the program will define.

Custom LCD Character using the "Character Box"

Character Box

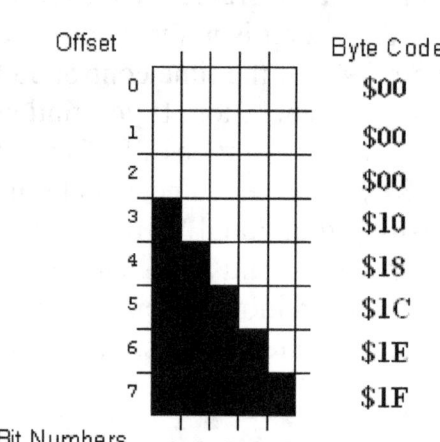

Figure 1: Custom LCD Character Using the "Character Box"

This custom character will be placed at memory location $00 of the LCD character memory. It then forms a small ramp that slopes downward.

Each row of the character has to be defined by a byte value. Since the characters are only 5 bits wide in size, the three most significant bits of the byte value are always zero. The highest bit value is the 5th bit (bit number 4).

The key is to set or clear the proper bits to form the character you want. For example, if you look at the fourth line of the character box (offset row value 3); the byte to the right shows $10 or binary %00010000. This makes the 5th block solid black, but the rest of the row clear. The next line sets two blocks black, by using byte $18 or binary %00011000, and the rest you can see in Figure 1. By setting

92

these bits and storing the data in the LCD's character memory, we have established a new custom character that the software can call.

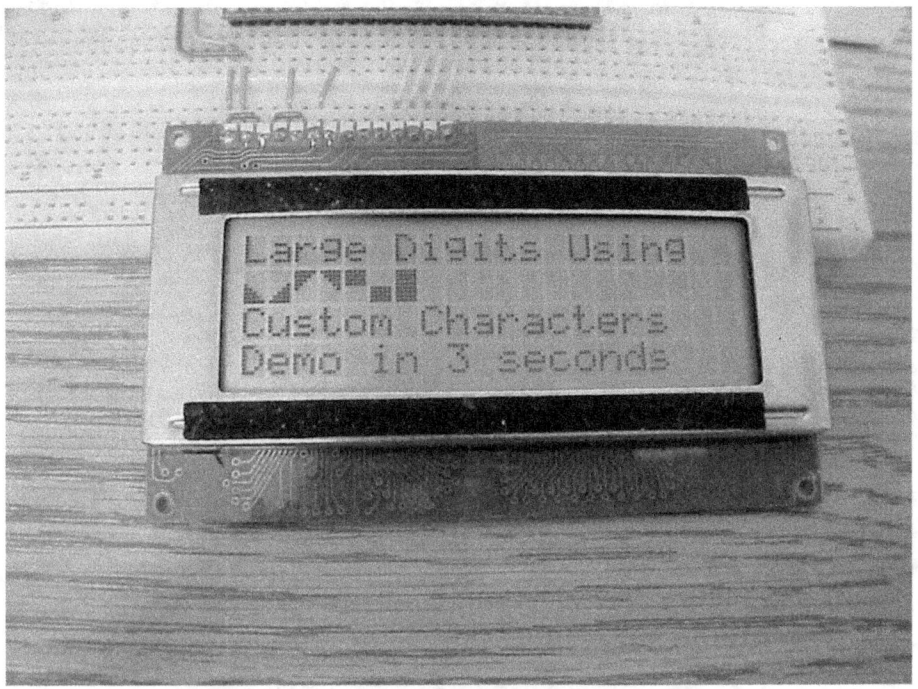

Figure 2: Custom LCD Characters

This doesn't stop us from displaying standard, single-line words on the LCD screen. It does allow us to also create large digits that span all four lines of a 4x20 LCD, by using the custom characters shown in Figure 2. It shows seven custom characters on line 2 of the LCD, but the eighth custom character is a blank so it doesn't appear. The main project software loop, discussed later, will create the hexadecimal number system and display them in sequence, starting with the numbers 0 – 9 and then A – F, all in a large-character format that spans all four lines. The number "1" is displayed in Figure 3 to illustrate what I'm trying to describe. Using large numbers like this makes it very easy to read from across the room.

Figure 3: An Example Large Character

Project Setup

If you remember the 2x16 LCD project that I did in a previous column, then you will find the 4x20 LCD has all the same connections. A great advantage to using LCD modules is their common connection system, which makes it easy to change from 2x16 to 4x20. Figure 4 shows the connections to the Ultimate OEM BasicATOM module. You can easily connect the LCD to a PIC16F876A directly, by following the connection names in the schematic.

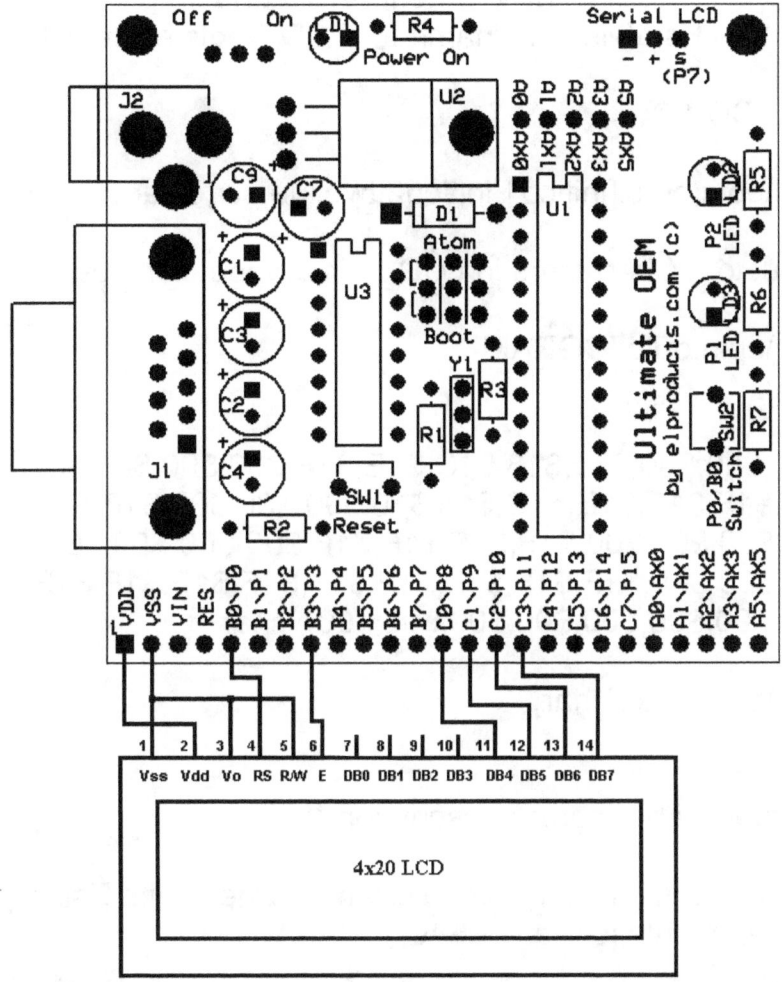

Figure 4: Connections to the Ultimate OEM BasicATOM Module

Software

This software listing is kind of long, but most of the code deals with setting up the LCD to display the large custom characters. In the code, you will notice the "|" pipe character at the end of several lines. This is for line continuation. This is a special character that the BasicATOM compiler recognizes as a continuation message. When the compiler sees that character, it knows the command line was too long for the editor window and continues on the next line. Setting up the characters takes a lot of space, so the line-continuation function is used often.

```
x var byte
char var byte
```

```
epin con 3          'Establish nickname for LCD enable pin
rspin con 0         'Establish nickname for LCD Register Select pin

' *** Initialize LCD ***
pause 500
lcdwrite rspin\epin,outc,[initlcd1,initlcd2,twoline,scr,clear,home]

'*** Create Custom Characters in LCD memory locations 0-7 ***

lcdwrite rspin\epin,outc,[CGRAM]
for x = 0 to 63

lookup x,[$00,$00,$00,$10,$18,$1C,$1E,$1F,$00,$00,$00,$01,|
$03,$07,$0F,$1F,$1F,$1E,$1C,$18,$10,$00,$00,$00,$1F,$0F,|
$07,$03,$01,$00,$00,$00,$1F,$1F,$1F,$1F,$00,$00,$00,$00,|
$00,$00,$00,$00,$1F,$1F,$1F,$1F,$1F,$1F,$1F,$1F,$1F,$1F,|
$1F,$1F,$00,$00,$00,$00,$00,$00,$00,$00], char

lcdwrite rspin\epin,outc,[char]
next

' *** Initial screen with program description ***
main
lcdwrite rspin\epin,outc,[clear,home,scrram,"Large Digits Using"]
lcdwrite rspin\epin,outc,[scrram + $40]

for x = 0 to 7
lcdwrite rspin\epin,outc,[x]
next
lcdwrite rspin\epin, outc,[scrram + $14, "Custom Characters"]
lcdwrite rspin\epin, outc,[scrram + $54, "Demo in 3 seconds"]
pause 3000

' *** LCD control code to display 0 - F large characters ****

' *** "0" character
lcdwrite rspin\epin,outc,[clear,home,scrram+$09,6,4,6,scrram+$49,|
6,7,6,scrram+$1D,6,7,6,scrram+$5D,6,5,6]
pause 1000
```

```
' *** "1" character
lcdwrite rspin\epin,outc,[clear,home,scrram+$09,1,6,scrram+$4A,|
6,scrram+$1E,6,scrram+$5D,5,6,5]
pause 1000


' *** "2" character
lcdwrite rspin\epin,outc,[clear,home,scrram+$09,4,4,6,scrram+$49,|
5,5,6,scrram+$1D,6,7,7,scrram+$5D,6,5,5]
pause 1000


' *** "3" character
lcdwrite rspin\epin,outc,[clear,home,scrram+$09,4,4,6,scrram+$49,|
5,5,6,scrram+$1D,7,7,6,scrram+$5D,5,5,6]
pause 1000


' *** "4" character
lcdwrite rspin\epin,outc,[clear,home,scrram+$09,6,7,6,scrram+$49,|
6,7,6,scrram+$1D,4,4,6,scrram+$5D,7,7,6]
pause 1000


' *** "5" character
lcdwrite rspin\epin,outc,[clear,home,scrram+$09,6,4,4,scrram+$49,|
6,5,5,scrram+$1D,7,7,6,scrram+$5D,5,5,6]
pause 1000


' *** "6" character
lcdwrite rspin\epin,outc,[clear,home,scrram+$09,6,4,4,scrram+$49,|
6,5,5,scrram+$1D,6,7,6,scrram+$5D,6,5,6]
pause 1000


' *** "7" character
lcdwrite rspin\epin,outc,[clear,home,scrram+$09,4,4,6,scrram+$49,|
7,7,6,scrram+$1D,7,7,6,scrram+$5D,7,7,6]
pause 1000


' *** "8" character
lcdwrite rspin\epin,outc,[clear,home,scrram+$09,6,4,6,scrram+$49,|
6,5,6,scrram+$1D,6,4,6,scrram+$5D,6,5,6]
pause 1000
```

' *** "9" character
lcdwrite rspin\epin,outc,[clear,home,scrram+$09,6,4,6,scrram+$49,|
6,5,6,scrram+$1D,7,7,6,scrram+$5D,7,7,6]
pause 1000

' *** "A" character
lcdwrite rspin\epin,outc,[clear,home,scrram+$09,6,4,6,scrram+$49,|
6,5,6,scrram+$1D,6,7,6,scrram+$5D,6,7,6]
pause 1000

' *** "b" character
lcdwrite rspin\epin,outc,[clear,home,scrram+$09,6,7,7,scrram+$49,|
6,5,5,scrram+$1D,6,7,6,scrram+$5D,6,5,6]
pause 1000

' *** "C" character
lcdwrite rspin\epin,outc,[clear,home,scrram+$09,6,4,4,scrram+$49,|
6,7,7,scrram+$1D,6,7,7,scrram+$5D,6,5,5]
pause 1000

' *** "d" character
lcdwrite rspin\epin,outc,[clear,home,scrram+$09,7,7,6,scrram+$49,|
5,5,6,scrram+$1D,6,7,6,scrram+$5D,6,5,6]
pause 1000

' *** "E" character
lcdwrite rspin\epin,outc,[clear,home,scrram+$09,6,4,4,scrram+$49,|
6,5,5,scrram+$1D,6,7,7,scrram+$5D,6,5,5]
pause 1000

' *** "F" character
lcdwrite rspin\epin,outc,[clear,home,scrram+$09,6,4,4,scrram+$49,|
6,5,5,scrram+$1D,6,7,7,scrram+$5D,6,7,7]
pause 1000

' **Final message from program before looping back to the top **

lcdwrite rspin\epin,outc,[clear,home,scrram,"Just imagine what",|
scrram+$40,"you can do!"]
pause 3000

goto main

How it Works
First, we establish a few variables and constants. The variables are just temporary storage locations labeled X and Char. The constants define the LCD "E" pin and "RS" pin.

```
x var byte
char var byte
epin con 3          'Establish nickname for LCD enable pin
rspin con 0         'Establish nickname for LCD Register Select pin
```

The program initializes the LCD by first waiting ½ second for it to warm up, and then it issues the LCDWRITE command to set it up as a two-line LCD. A 4x20 LCD is really a two-line by 40 character LCD, with line 1 broken up between lines 1 and 3, and line 2 broken up between lines 2 and 4, to form a 4x20 LCD.

```
' *** Initialize LCD ***
pause 500
lcdwrite rspin\epin,outc,[initlcd1,initlcd2,twoline,scr,clear,home]
```

The next section is the heart of this program. In this block of code, the custom characters are created and stored in the LCD character memory locations 0 thru 7. Each character takes 8 bytes of data, for a total of 64 bytes (8 characters times 8 bytes).

To do this, we first have to point to location zero of the Character RAM. We do this with the LCDWRITE command, again by sending the "CGRAM" pointer. We don't have to add an address value, since it defaults to the zero or first location.

```
'*** Create Custom Characters in LCD memory locations 0-7 ***

lcdwrite rspin\epin,outc,[CGRAM]
```

Now, we send the custom characters to the LCD character RAM by using a FOR-NEXT loop and the LOOKUP command. The FOR-NEXT loop counts from 0 to 63, for a total of 64 loops, and it defaults to stepping one count per loop. The variable "x" stores the present loop count value. The LOOKUP command then takes the value of "x" and jumps that many places, reads the byte value and stores it in the "char" variable. For example, lets assume x = 5 or the 6th time through the loop since the count starts at zero. The value of "char" will equal $1C, since it is the sixth value listed.

```
for x = 0 to 63

lookup x,[$00,$00,$00,$10,$18,$1C,$1E,$1F,$00,$00,$00,$01,|
$03,$07,$0F,$1F,$1F,$1E,$1C,$18,$10,$00,$00,$00,$1F,$0F,|
$07,$03,$01,$00,$00,$00,$1F,$1F,$1F,$1F,$00,$00,$00,$00,|
$00,$00,$00,$00,$1F,$1F,$1F,$1F,$1F,$1F,$1F,$1F,$1F,|
$1F,$1F,$00,$00,$00,$00,$00,$00,$00,$00], char

lcdwrite rspin\epin,outc,[char]
next
```

After the code above executes, the custom characters are now in the LCD character-generator memory. The program can now call them to create the large characters on the LCD. The "main" label starts the central program loop. In the section below "main", we use the LCDWRITE command to display a description of what this program will do, as shown in Figure 2. We display "Large Digits Using" by using the LCDWRITE command.

```
' *** Initial screen with program description ***
main
lcdwrite rspin\epin,outc,[clear,home,scrram,"Large Digits Using"]
lcdwrite rspin\epin,outc,[scrram + $40]
```

This next section will call up the custom characters just created, one at a time, using a FOR-NEXT loop, and will display them using the LCDWRITE command. The variable "x" holds a value from 0 to 7. LCDWRITE directs the LCD to display characters 0 thru 7. See how easy it is to display custom characters, once they are created?

```
for x = 0 to 7
lcdwrite rspin\epin,outc,[x]
next
```

We finish this block of code by displaying "Custom Characters" and "Demo in 3 seconds" to the display lines 3 and 4. SCRRAM +$14 is the beginning of line 3, and SCRRAM +$54 is the beginning of line 4.

```
lcdwrite rspin\epin, outc,[scrram + $14, "Custom Characters"]
lcdwrite rspin\epin, outc,[scrram + $54, "Demo in 3 seconds"]
```

Finally, we pause 3 seconds so you can read the display and then move on to the next section.

```
pause 3000
```

From here, the program creates the custom large characters using the custom 5x8 characters stored in CGRAM. I'll just describe the digit "1", shown in Figure 3, but all the other large character sections below operate in the same manner.

The #1 character is created by placing custom characters 1 and 6 on line one, character 6 on line two, character 6 on line three, and characters 5,6 and 5 on line four. We then pause one second, so the digit can be read. The scrram is offset with values that center the "1" on the LCD.

```
' *** "1" character
lcdwrite rspin\epin,outc,[clear,home,scrram+$09,1,6,scrram+$4A,|
6,scrram+$1E,6,scrram+$5D,5,6,5]
pause 1000
```

The rest of the large digits (0 thru F) are created in a similar fashion. Each is displayed, and then a final message is displayed. The final section of code displays, "Just imagine what you can do."

```
' *** Final message from program before looping back to the top ***

lcdwrite rspin\epin,outc,[clear,home,scrram,"Just imagine what",|
scrram+$40,"you can do!"]
pause 3000
goto main
```

With this custom character method, you can create just about anything on an LCD screen.

Next Steps

The projects that can result from this are endless. The thing to remember is, nothing stops you from redefining the custom characters in the middle of the program. For example, let's say you want to display large characters, initially, and then later in the program want to create an animation using different custom characters.

After completing the large custom number characters, clear the LCD screen and then load new custom characters in CGRAM locations 0 - 7. From these new characters, you can create the animation. Since the custom characters load in CGRAM so fast, the person watching the display just notices a frame change from words to large digits to animation. I've seen custom characters that had the old Pacman character eating dots across the screen.

Conclusion

By no means am I declaring that I invented this custom-character method. In fact, there are numerous sites on the Internet that refer to creating custom characters on an LCD. Scott Edwards even incorporates this type of large-character generation into his PIC MCU-driven serial LCD modules. If you are really creative, you can probably create a whole animated cartoon on the LCD by constantly changing the custom characters. It will potentially take a lot of memory, but most graphic programs do.

I once again use the BasicATOM chip, because of the simplicity of the software and the low cost for any reader that wants to follow along by doing the projects. I have received many emails from readers asking me to pick a platform and stick with it. Because Basic Micro (creator of the BasicATOM) also has a Basic compiler, which is called the MBasic Professional compiler, this platform offers the reader the option to start cheap with the BasicATOM software, via a free download and a $20 BasicATOM chip (which should be cheaper by the time you read this). Eventually, most people will move to programming blank PIC MCUs, to save money on larger-volume projects.

Here's a tip for readers; there is a book available at the Nuts & Volts Web site called "Programming the PIC Microcontroller in MBasic" by Jack Smith (this has a similar title and the exact same publisher as my PICBASIC compiler book). Jack does a great job of detailing how to use the PIC16F877A MCU with MBasic Pro. Best of all—his book's CD includes a free version of the MBasic Pro compiler, which is limited to working with the PIC16F876(A). This offers you the option to take a 28-pin BasicATOM chip design and pretty easily move it to PIC16F876A—all for the cost of a book and a programmer (to load the .hex file into the PIC16F876A). Check it out.

Email me with your comments at chuck@elproducts.com, and thanks for all of the feedback I continue to get. I do like reading the feedback and try to respond to all of the emails as quickly as possible. See you next month.

October 2007 - PICkit 2 Starter Kit vs. PICkit 2 Debug Express

If you've been reading this column on a regular basis you will, no doubt, have realized that I'm always on the lookout for a better and lower-cost starter package, or at least programming solution. I've recommended Microchip Technology's PICkit™ 2 Starter Kit in the past, and have also mentioned its cousin—the PICkit 2 Debug Express. I've received several emails asking me to cover the PICkit 2 Debug Express in more detail, since there is apparently some confusion over which version of the PICkit 2 to buy. What I found out is the PICkit 2 Starter Kit and Debug Express are identical except for the demo board included so the choice gets much easier to make.

PICkit 2 Starter Kit vs. PICkit 2 Debug Express

If you visit the Microchip Web site and see both the PICkit 2 Starter Kit (Part Number: DV164120) and the PICkit 2 Debug Express (Part Number: DV164121), you may notice that they look very similar and they both cost $49.99. They look similar because the programmer/debugger electronics are identical. The hardware is exactly the same with possibly different a firmware level installed. In fact, Microchip makes the firmware free to download, so you can update the firmware version on either package you purchase to the latest identical level.
The software to control the PICkit 2 comes in two varieties; a standalone application for programming and Microchip's MPLAB® Integrated Development (IDE) for programming/debugging. To use it as a programmer or programmer/debugger—in MPLAB you choose that through the MPLAB IDE's programmer or debugger menu option (see Figure 1).

Note that earlier I stated "possibly" different firmware levels. The reason I say this is because the firmware level used by the MPLAB interface and the firmware level used by the standalone interface were at different release levels. The standalone interface doesn't offer debugging so I assumed the Debug Express only worked with the MPLAB interface. If you took a PICkit 2 and used it with the standalone interface and then later switched to the MPLAB interface, MPLAB would automatically update the firmware. This made me think there was a different firmware used for the PICkit2 and the DEBUG Express but that was not true. They were just different release levels. All this gets corrected with MPLAB version 7.62

which was just released on the Microchip website as I write this column. Both the standalone application and the MPLAB IDE will update the PICkit 2 to identical firmware version 2.10. So this means both the PICkit 2 Starter Package and the PICkit 2 Debug Express are identical except for the demo board. Deciding which one to buy is simplified down to which demo board do you want.

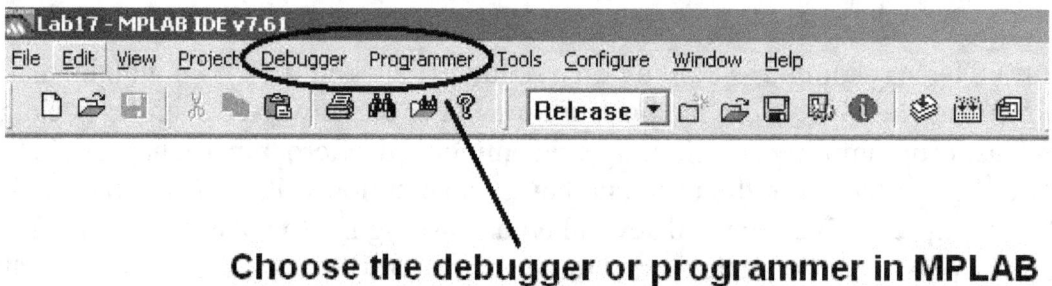

Choose the debugger or programmer in MPLAB

Figure 1: MPLAB IDE Menu Options

Firmware
Before I get into the different demo board choices you may be wondering what firmware is? I'm really not sure where the term came from, but firmware is just another term for the software that controls the hardware. The difference is that firmware is written, compiled and ready to load in the hardware, while software is typically the source code that can be modified or customized for your hardware or application—but then you have to compile it and create the .hex file. In the case of the PICkit 2 Starter Kit and PICkit 2 Debug Express, the hardware is identical and so is the firmware . Updating the operation to support the latest PIC MCU's just requires a new firmware download.

Many third-party PIC® microcontroller (MCU) programmers work this same way. They have a control MCU inside (typically another PIC MCU) that receives the .hex file and then sends it to a socketed, blank PIC MCU to be programmed. The software that runs the programmer's control MCU is called the firmware, and it can be updated as new PIC MCUs are released. This is also how the PICkit 2 Microcontroller Programmer hardware functions. At the heart of the PICkit 2 Microcontroller Programmer is a PIC18F2550, which has the USB drivers built in. The PIC18F2550 uses a bootloader to self program the firmware via its USB port. One of the nice things about the PICkit 2 Microcontroller Programmer is that the 5V USB line is passed through to the application, so you can also power your circuit from the PICkit 2 Microcontroller Programmer's programming connector. You can get the full PICkit 2 Microcontroller Programmer schematic in the user guide that comes on the included CD, or you can download it from the microchip.com/PICkit Web site.

Figure 2: PICkit 2 Starter Kit

PICkit 2 Starter Kit

So why are there two different starter packages, if they both come with the same programmer/debugger? Since the PICkit 2 Starter Kit and the PICkit 2 Debug Express use the same interface, Microchip just made it convenient by selling two variations of the demo board—to make it either a programmer or a programmer/debugger, right out of the box. Because of this commonality, it makes sense that some beginners or even experienced users might wonder which unit to purchase. It really depends on what you plan to do with it. If you just want to use the 20-pin, 14-pin and 8-pin PIC MCUs, then the PICkit 2 Starter Kit for $49.99 is the way to go (see Figure 2). The included PICkit 2 Low Pin Count Demo Board has a 20-pin socket and several simple components already soldered to the board—including LEDs, a potentiometer and a momentary switch. The 20-, 14- and 8-pin PIC MCUs all share the same programming connections on the top 8 pins, so the 20-pin socket accepts all three package sizes for programming. The PICkit 2 Starter Kit also includes a PIC16F690, to get you started.

The PICkit 2 Starter Kit is listed as part number DV164120, and can be ordered directly from www.MicrochipDirect.com. This is clearly a great choice if you just want to write code and then program. The problem arises when you decide you want to use the MPLAB IDE debugging feature, which allows you to single step through your code, run in animate mode or run to a breakpoint. The 20-, 14- and 8-pin PIC MCUs do not have the debug executive silicon inside, because of the die size limitations. To debug with these parts, you need an adapter (Figure 3) that matches the part you want to debug. The adapter has a custom chip with the debug executive, and the rest of the chip consists of "normal" silicon in a larger package with extra pins for the debugger connection—so you can debug code for these smaller PIC MCUs. This adds more cost, but the adapters are only around $25 - $35.

So, this leads us to the PICkit 2 Debug Express. If you really want to debug your code and you are using a PIC MCU with the debug silicon on it, which includes almost every 18-, 28-, 40-pin and larger PIC MCU, then you might want to consider using the PICkit 2 Debug Express.

Figure 3: Debug Adapter

PICkit 2 Debug Express
The PICkit 2 Debug Express is the better choice to start with, if you are just learning about PIC MCUs or are developing a design that needs a lot of I/O. The PICkit 2 Debug Express comes with a development board that has a 44-pin, surface-mount PIC16F887 soldered to it. The PIC16F887 is a great part for development, because it has most of the popular features designers want in their PIC MCUs; and it has the debug executive silicon built in, so it can be debugged in-circuit. As I've discussed in previous articles, a debugger lets you run your code in animate mode, single step or run to a set breakpoint. This can be quite handy when you get into bigger programs that are hard to follow.

Sometimes a simple typo can be accepted by the compiler as a good command, but it's doing something different than what you expect. That type of error can be difficult to find, without a debugger helping you to narrow the search. That is really what a debugger does for you. You can set a breakpoint just above the section of code that you know isn't working properly. Then, after the breakpoint you can single step through the code to see where the operation may be running off track.

Microchip also offers the MPLAB ICD2 debugger, which supports all of the PIC MCU families. However, the MPLAB ICD 2 Module (Part Number: DV164005) costs $159.99, so the PICkit 2 Debug Express is a lower-cost alternative in a very small package. Figure 4 shows the PICkit 2 Debug Express.

Figure 4: PICkit 2 Debug Express

The PICkit 2 in Debug mode previously did not support - many parts for debugging , but that changes with the release of MPLAB 7.52. Most of the 16F and 18F PIC MCU's with the debug executive are supported.

One of the limitations of the PICkit 2 Debug Express is the surface-mount design of its included Demo Board. Because the PIC16F887 is surface mounted to the demo board, this package is not the best option if you want to program another PIC16F887 by plugging a blank part into a socket. This makes the PICkit 2 Debug Express a little less useful to the hobbyist. So, if you want a programmer/debugger that comes with a demo board that has a DIP socket and accepts PIC MCUs that have the debug executive silicon, then I have another option for you.

My Choice

Microchip offers several different PICkit 2 Demo Boards, separately, that plug right into the PICkit 2 Microcontroller Programmer. You can even buy different versions of the demo boards, in addition to the ones included with the starter packages. These extra boards are sold in packs of three, with one of the three populated just like the starter kit board and two more that are completely blank—ready for you to solder onto your own design.

One of the demo boards offered by Microchip has a 28-pin DIP socket and includes a PIC16F886 on the completed board. This PICkit 2 28-pin Demo Board (Part Number: DM164120-3) is actually part of Microchip's PICkit Serial Analyzer, which I will save for another article . The reason I mention this demo board is because it fits the format I described above. The board will accept any 28-pin PIC MCU in a 0.300 wide DIP package, and has the PICkit 2 programming/debugging header for plugging into the PICkit 2 Microcontroller Programmer. This opens up lots of opportunities.

As you know, I focus on the BASIC language for many of this column's projects because it's an easy language for the beginner to use. If you want to start with BASIC, then my idea is to create your own custom starter package by first buying the standalone PICkit 2 Microcontroller Programmer, for $34.99. This saves a little money to purchase the demo board of your choice. The USB cable comes with the programmer and all of the starter kit software is on the included CD. The PICkit 2 Microcontroller Programmer, alone, can be purchased under the part number PG164120 at www.MicrochipDirect.com (see Figure 5).

The PICkit 2 28-pin Demo Board can be purchased at the same time, under the part number DM164120-3, for $24.99. The reason this combination works best for the BASIC programmer is because the PICkit 2 28-pin Demo Board (see Figure 5) can accept a PIC16F876A. Then, you can use the free sample version of microEngineering Labs' PICBASIC PRO compiler, which supports the PIC16F876A, to create your code. (You can download the sample version from melabs.com.)

The sample version of the PICBASIC PRO compiler can also be run in Microchip's MPLAB IDE, which comes on the PICkit 2 CD or can be downloaded for free from Microchip's Web site. More information on how to set this up is at: melabs.com/support/mplab.htm. Using this PICBASIC PRO compiler within the MPLAB IDE provides all of the debugging features, plus the one-touch compile and program features of the MPLAB IDE/PICkit 2 setup.

There is a catch to my custom setup, though. You will have to add a crystal or resonator to the board, since the PIC16F886 doesn't need an external oscillator but the PIC16F876A does. I suggest that you use a 3-pin SIP socket, so you can plug in various crystal/resonator speeds. The development board traces are already designed to accept an external crystal/resonator, so this modification is easy for the beginner to accomplish. The total cost of this custom setup is $59.98 or about $10 more than the standard starter packages (but you also get two blank PICkit 2 28-pin Demo Boards for future designs, to justify the extra $10).

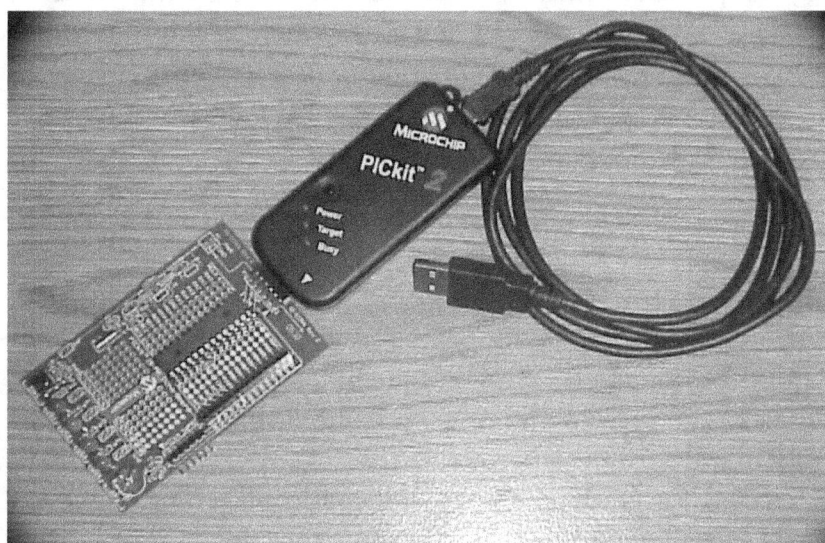

Figure 5: Chuck's Custom Starter Package

To use the PIC16F876A in debug mode, you'll need MPLAB 7.61 or later. Hopefully, this version or later will be on the included CD—but if an earlier version of the MPLAB IDE is included, you can download the latest version of the MPLAB IDE for free from microchip.com/mplab.

To summarize my recommendations, you can use the sample version of the PICBASIC PRO compiler and compile, program or debug from within the MPLAB IDE—using the PICkit 2 Microcontroller Programmer in programmer mode or programmer/debugger mode. You can then program more blank 28-pin DIP package PIC MCUs, as you create more projects with this great starter package. Getting all this for $58.98, plus tax and shipping, seems like a great deal to me.

More Options

Now, it doesn't stop there. If you read last month's column, you also might remember that I mentioned Jack Smith's book "Programming the PIC Microcontroller with MBasic", which includes a full working version of Basic Micro's MBasic Professional compiler—limited to compiling the PIC16F876 or PIC16F876A. MBasic will not run within the MPLAB IDE, but you can still use the PICkit 2 Microcontroller Programmer with the MBasic Professional compiler by interfacing the standalone version of the PICkit 2 PC software for programming PIC MCUs outside of the MPLAB IDE.

You do that by simply importing the .hex file generated by the MBasic Pro compiler from Jack's book into the PICkit 2 software. When I last checked, Jack's book was selling for $62.95 on amazon.com, but that is cheaper than a full-featured compiler. It's another low-cost option for the hobbyist to get started with programming PIC MCUs.

PICBASIC PRO

If you do choose to use PICBASIC PRO then, after you get comfortable with the PICBASIC PRO sample version, you might be ready to shell out the $250 for the full PICBASIC PRO. This will give you the ability to program the other 28-pin PIC MCUs, such as the PIC16F886 or even some of the 28-pin PIC18F parts—which also plug into the 28 pin board. The full version of PICBASIC PRO supports most of 8-Bit PIC MCU's from the small PIC10F up to the PIC18F family of parts. Later, you can add the 20-pin development board to your collection and program the 20-, 14- and 8-pin PIC MCUs with PICBASIC PRO, as those are also supported. You can even prove out your design with a 28-pin part, using the debugger, and then recompile it for a smaller part using the programmer and the 20-pin board. I've successfully done this, but you really have to understand the differences between the two parts to do this properly.

I'm sure new demo boards will be released in the future, because the PICkit 2 Microcontroller Programmer is so handy to use. I have heard about a few third-party boards that are also in the works for the programmer. If BASIC is not your language of choice and you prefer to program in C, the HI-TECH PICC-Lite compiler is also included on the PICkit 2 CD. Unfortunately, the PICC-Lite compiler doesn't support any 28-pin PIC MCU parts. It does support several smaller parts that fit the 20-, 14- or 8-pin packages. The PICC-Lite compiler also supports the PIC16F887, which is the surface-mount development board part. The PICC-Lite compiler can also be run inside the MPLAB IDE, so you get a professional style setup for a very low price. The CCS PCM™ Midrange C Compiler Demo for PIC16F887 is also now part of the software included on the CD to give you another choice.

The sample versions of the various compilers I mentioned don't support the PIC18F parts, but you're in luck here as well. Microchip does offer a student version of their MPLAB C18 compiler, which you can download for free at microchip.com/c18. The only limitation of this free version is that the compiler loses some optimizations after 60 days, but that's not a big deal for most beginners. Besides, this compiler supports all of the PIC18F MCUs. This is another option for the beginner interested in C programming, and this will work with my makeshift 28-pin starter package idea.

Conclusion

I hope this wasn't too confusing, and that I successfully clarified the differences between the two PICkit 2 starter packages for you. I added a PICkit 2 header to my Ultimate OEM board that has a 28-pin socket, to see how well it would work. I was extremely happy with the setup, since I now have a great connection between the MPLAB IDE and my breadboard module's development setup. This allows me to do just what I described above, without having to buy the PICkit 2 28-pin Demo Board. It works great.

In conclusion, I recommend that you make your own starter package with a PICkit 2 Microcontroller Programmer, a PICkit 2 28-pin Demo Board, a sample version of the PICBASIC PRO complier and Microchip's free MPLAB IDE, to get started in programming PIC MCUs. You'll also have a tool set that can grow with your programming skills, without breaking the bank account. Programming PIC MCUs just gets better and better, doesn't it?

Keep on giving me your feedback. I read it all. Send to: chuck@elproducts.com.

114

November 2007 - PICBASIC PRO, EXCEL and the BEGINNER

Good things come to those who wait is a common saying but the real question is how long do I have to wait? Some would say it all depends on how you define a "good thing". In my case, several years ago I asked the guys at microEngineering Labs for a sample version of their PICBASIC PRO compiler. You might even say I was a pest. It has always been my opinion that a demo version was the best way for a beginner to learn for very little cost. I also wanted a demo version for selfish reasons, so I could include it in educational training books that I'm finally getting a chance to work on. Eventually microEngineering Labs released a sample version. It was worth the wait. I've used here in earlier installments of this column and have always been grateful for that demo.

I ran into the owner of microEngineering Labs at this years Microchip's Masters Conference and I told him I get a lot of reader feedback both from this column and elsewhere that many beginners would like to work with the eight pin PIC MCU's. I asked him if he would consider adding an eight pin part to the sample version for those just getting started. It was the first and only time I asked and to my surprise he delivered it to the world only a couple of weeks later. In fact microEngineering Labs did better than that, they also added the 14 pin 16F688 and the 20 pin 16F690 which is the same part used in the PICkit 2 Starter Package I talked about last month. You might be thinking "who cares", why is this worth mentioning? In a general sense it isn't that big of a deal to some but to many beginners out there that still want a simple platform to get started with without making a big investment it's a huge bonus.

Many of the comments state they like what they see in PICBASIC PRO but are concerned that they will not understand programming well enough to make it worth investing $250 up front. Well now the beginner can use an off the shelf eight pin, 14 pin or 20 pin PIC MCU and a simple programmer to create some very interesting projects within the 31 command limit set by the sample version of PICBASIC PRO. The big deal in my mind is when a beginner takes this path, they can successfully program a PIC MCU and get into that comfort zone of knowing they can do it. The only thing left to do after that is decide when they want or need to spend the $250 to get the full version of PICBASIC PRO.

Starting out with these smaller parts has the advantage of using the latest Microchip PIC MCU technology. All three of these parts have internal oscillators and internal MCLR pull-up resistors. That means after it's programmed all you need is 5v and ground to make the chip function. In fact you can even run them from less than 5v. Using a few AA batteries will often work so the intimidation factor is even less. These parts also share a common pin-out for the upper eight pins as shown in Figure 1. This makes using these easier because they can all share the same development board. I think these are some of the greatest parts for a beginner to use and now you can program them with the free PICBASIC PRO sample version from www.melabs.com.

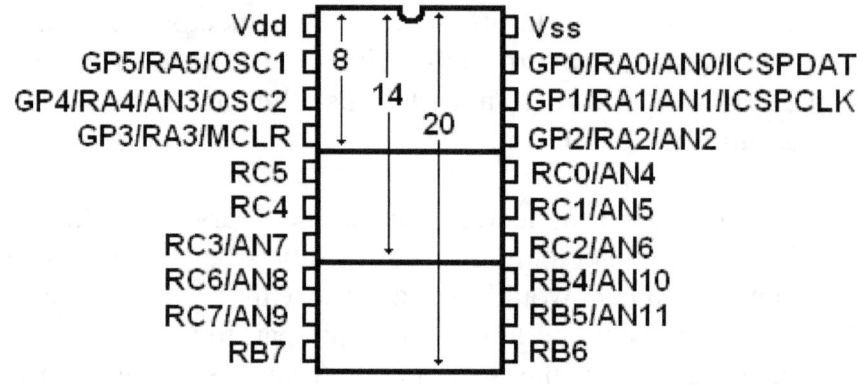

Figure 1 – Common Pin-Out

PICkit 2 USART Tool

As I mentioned, with the sample version you still need a hardware programmer to load the code into the chip. I've talked about many of them in the past including my own design which is now only available at BeginnerElectronics.com since my site no longer sells products other than my books. I decided I wanted to quickly test out this new demo version for this article but in addition to that I wanted to try out the new PICkit 2 standalone programming interface version 2.40 that runs by itself outside of Microchip's MPLAB (shown in Figure 2). One of the features of this new version I wanted to test out is a UART tool under the "Tools" menu item.

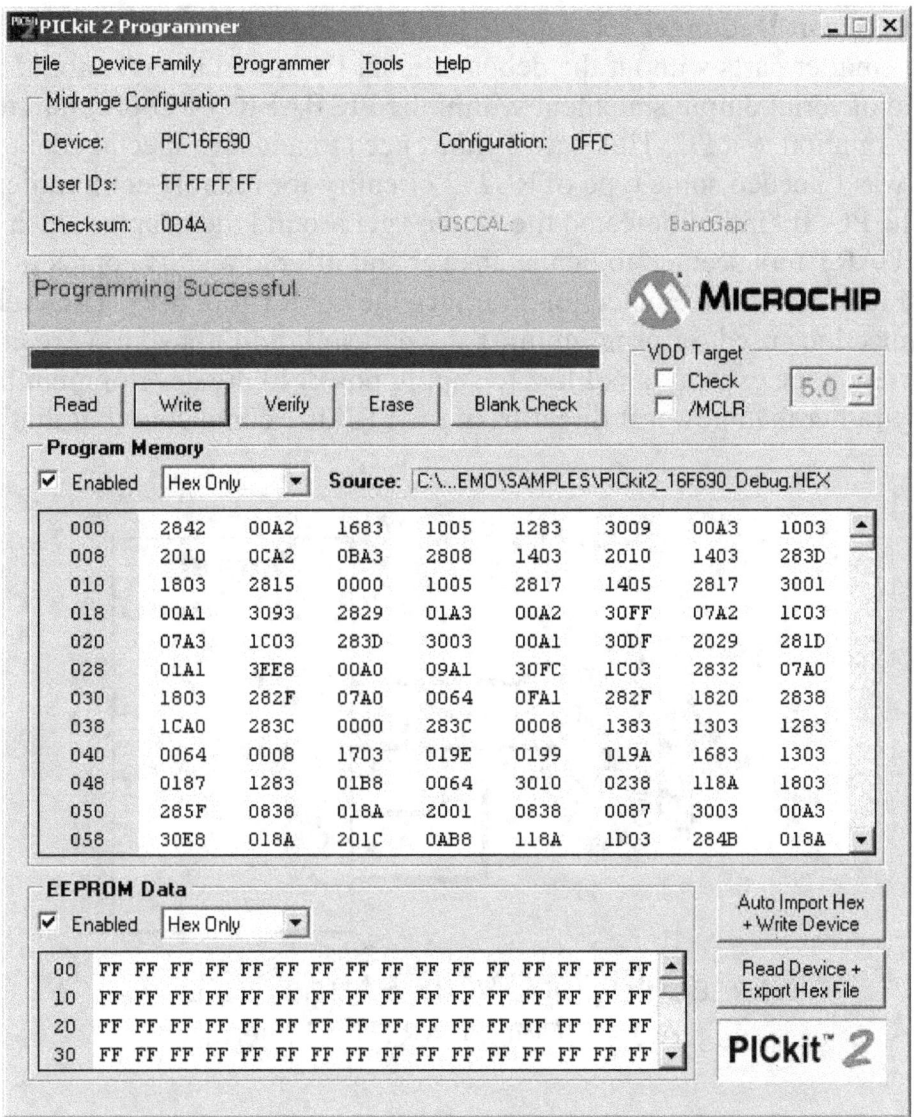

Figure 2 – PICkit 2 Standalone Programmer Interface

I could not have picked a better time for this new demo version of PICBASIC PRO to be released so I could test it with this new PICkit 2 interface. Its killing two birds with one stone (another common saying). You see one of the limitations of these smaller micros is the package size doesn't leave enough space for the debug silicon. If you want to debug in-circuit with Microchip's typical debug tools then you have to use a special adapter. With this new software interface I had hoped to create a simple alternative in- circuit debugger.

Serial In-Circuit Debugger

For these smaller parts without the debug silicon, I've often thought about using some kind of serial output statement within the PICBASIC PRO program to send status information to a PC. This way I didn't need to add any special adapter. The problem was I needed some type of RS232 circuitry for reliable communication back to the PC. It kind of defeated the whole "get around the adapter" idea. The PICkit 2 UART tool seemed to be the answer and offered a USB connection as well. Figure 3 shows the connection interface the UART tool uses. These are the same connections used for programming the part so I didn't have to wire anything up separately. One exception is I had to supply power to the development board externally rather than power it direct from the PICkit 2 connection but that's not tough.

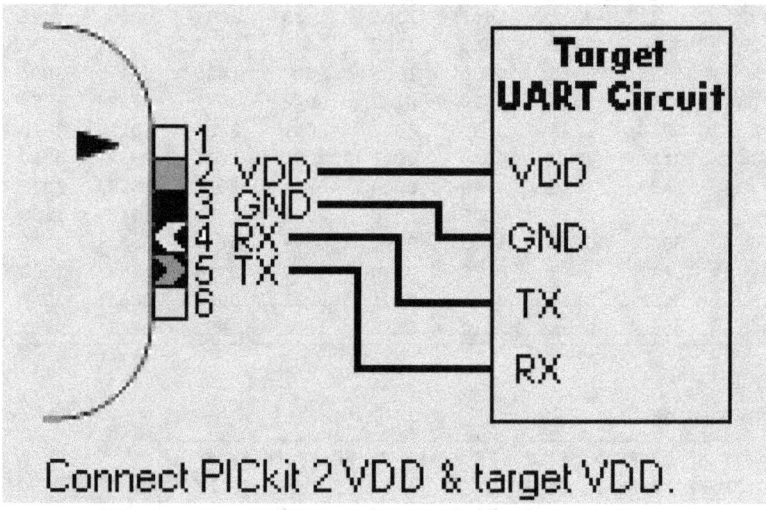

Figure 3 – PICkit 2 UART Tool Interface

For this test I decided to use the Low Pin Count Demo board that's included with the PICkit 2 Starter Package I covered in last month's article. I also decided to use a PIC16F690 (since I didn't have any 12F683's or 16F688's in my basement lab). On the 16F690 (and 16F688) the RA0 pin (GP0 on the 12F683) is located at the same pin in the 20 pin socket so this technique should work on any of these parts. By sending data serially out the connection to the PICkit 2 connector, the PICkit 2 will convert it to USB format and then the GUI software will display the information. This means a simple serial command inside the PICBASIC PRO program can send back the status of a variable or send some kind of message to the PC. This seemed easy enough to do with a SEROUT command but then I remembered that PICBASIC PRO has a DEBUG command built it just for this type of communication. My path was set to test this idea and listing 1 shows the resulting program.

LISTING 1

```
DEFINE DEBUG_REG PORTA          ' Set DEBUG Port to PORTA
DEFINE DEBUG_BIT 0              ' Use pin A0 of PORTA for DEBUG
DEFINE DEBUG_BAUD 2400         ' Set Baud rate to 2400
DEFINE DEBUG_MODE 0           ' Communicate in True Mode


ANSEL =    0                   ' Intialize A/D ports off
CM1CON0 = 0                    ' Initialize Comparator 1 off
CM2CON0 = 0                    ' Initialize Coparator 2 off
TRISC = 0                      ' All PORTC set as outputs

LEDs  var  byte                ' Establish counter variable

loop:

for LEDs = 0 to 15             ' Step through Binary count from 0 to 15

'       debug LEDs ' Monitor variable value through PICkit 2 Window (Hex Mode)
        debug #LEDs,$0D, $0A ' For Data Log to Excel File mode (ASCII Mode)

        PORTC = LEDs     ' Make LEDs on board match Variable value
        pause 1000       ' Delay 1 second to watch LEDs

next

Goto loop        ' Loop through this forever
```

The program is very easy to follow but I'll step through some of the details. At the top are the setup DEFINE's for the DEBUG command. Some of these are default but I set them up anyway. I make the DEBUG command communicate through PORTA pin RA0 at 2400 baud in True mode. True mode means the PICkit 2 has an inverter built in so true RS232 signal format is required (though it ends up being USB format in the end).

```
DEFINE DEBUG_REG PORTA          ' Set DEBUG Port to PORTA
DEFINE DEBUG_BIT 0              ' Use pin A0 of PORTA for DEBUG
DEFINE DEBUG_BAUD 2400         ' Set Baud rate to 2400
```

```
DEFINE DEBUG_MODE 0              ' Communicate in True Mode
```

This next section is dedicated to setting up the PIC16F690. Port A defaults to analog inputs so I have to clear the ANSEL register to turn off the A/D ports and make Port A all digital. The PIC16F690 also has comparators built in that I wanted to make sure they were turned off. Clearing the CM1CON0 and CM1CON1 registers takes care of this. Finally Port C TRIS register is cleared to make all of Port C output pins to drive LEDs. Below this I create a variable named "LEDs" to hold the LEDs display state.

```
ANSEL =     0               ' Intialize A/D ports off
CM1CON0 = 0                 ' Initialize Comparator 1 off
CM2CON0 = 0                 ' Initialize Comparator 2 off
TRISC = 0                   ' All PORTC set as outputs

LEDs  var  byte             ' Establish counter variable
```

The last part of the program is a simple For-Next loop that increments the variable "LEDs" from 0 to 15 and displays the value as a binary display on the four demo board LEDs.

```
loop:

for LEDs = 0 to 15          ' Step through Binary count from 0 to 15

'      debug LEDs  ' Monitor variable value through PICkit 2 Window (Hex Mode)
       debug #LEDs,$0D, $0A ' For Data Log to Excel File mode (ASCII Mode)

       PORTC = LEDs     ' Make LEDs on board match Variable value
       pause 1000       ' Delay 1 second to watch LEDs

next

Goto loop       ' Loop through this forever
```

In the middle of the For-Next loop is a DEBUG command line. In fact there are two, but one in commented out. I did two of these for a purpose. The simpler one sends the state of the variable "LEDs" as a raw data value. The UART tool will display it as a hex value if I select that button. Figure 4 shows the UART tool screen and I highlight some of the features including the HEX vs. ASCII selection button in the upper right corner. The data received is shown on the left of the screen in HEX mode.

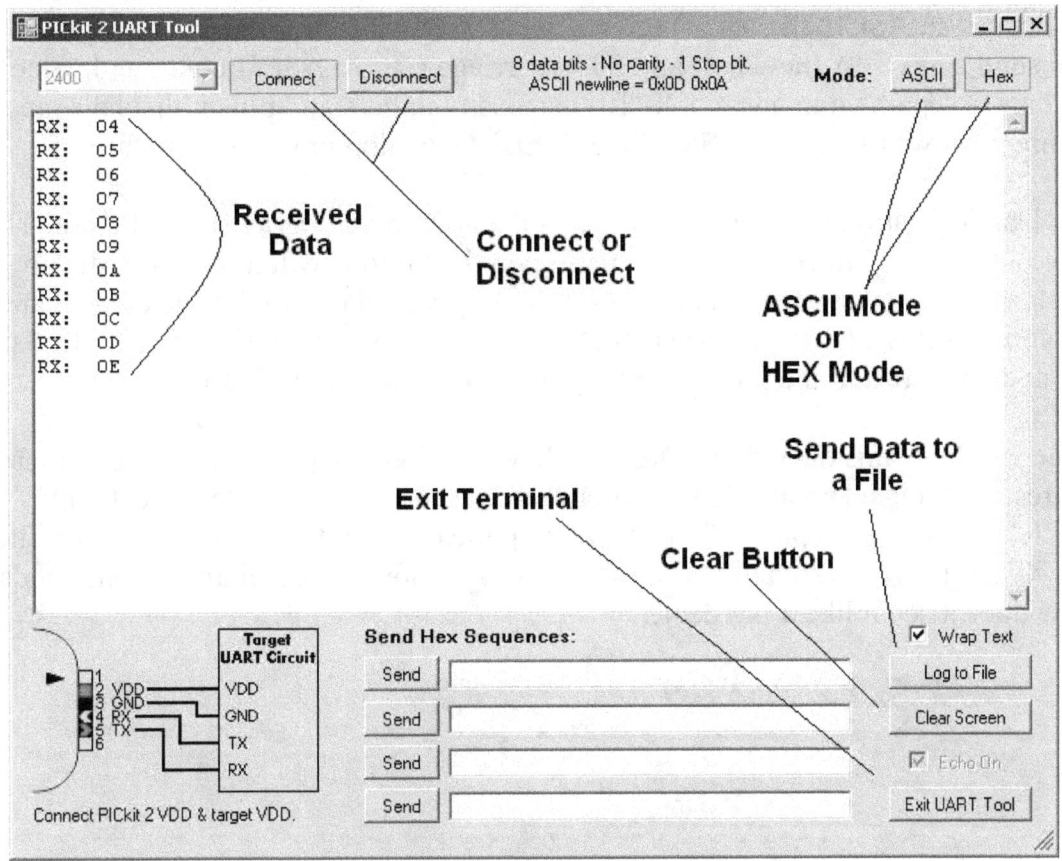

Figure 4 – UART Tool Window

Notice how the prefix "RX:" is put in front of the data. The UART tool does this automatically. As you can see from Figure 4, the value of "LEDs" is displayed so I can monitor the variable and watch it change. Now this data being displayed could also be a register value or a timer value or it could simply be a special code that indicates where the program is at if it suddenly starts doing something you don't expect. Knowing where to look for a coding error is half the intent of in-circuit debugging. So far this new method of debugging was working great. Then it got better.

Sending Data to EXCEL

One of the great features I discovered in this UART tool is the ability to easily store data in a file that can be loaded into Microsoft EXCEL. The "Log to File" button takes care of this. I didn't want the "RX:" included in the EXCEL file, just the data, so the second DEBUG command line takes care of this. That line sends the data as an ASCII character because the "#" sign is in front of the variable. I also send a hex $0A (newline) and $0D (carriage return) ASCII command value so the data is separated in a way EXCEL can easily understand plus it displays nicely. Figure 5 shows the EXCEL file of this DEBUG datalogging to file method.

The Datalog button turns green when it is datalogging and gray when it's not. This makes it real easy to control. Just clicking on the button switches it from datalog off to datalog on and back again. You have to give a name to the file created and I created "test.xls". By using the suffix ".xls" this allowed EXCEL to easily find this file and convert it to the proper format when I opened it in EXCEL.

Once you have this data in an EXCEL file you can easily plot it and do all kinds of interesting things. Having a small micro read a sensor and storing the data into EEPROM then at the push of a button dump the data out the PICkit 2 to a log file in EXCEL format is something that can easily be done in less than 31 commands. Now does it seem like a big deal?

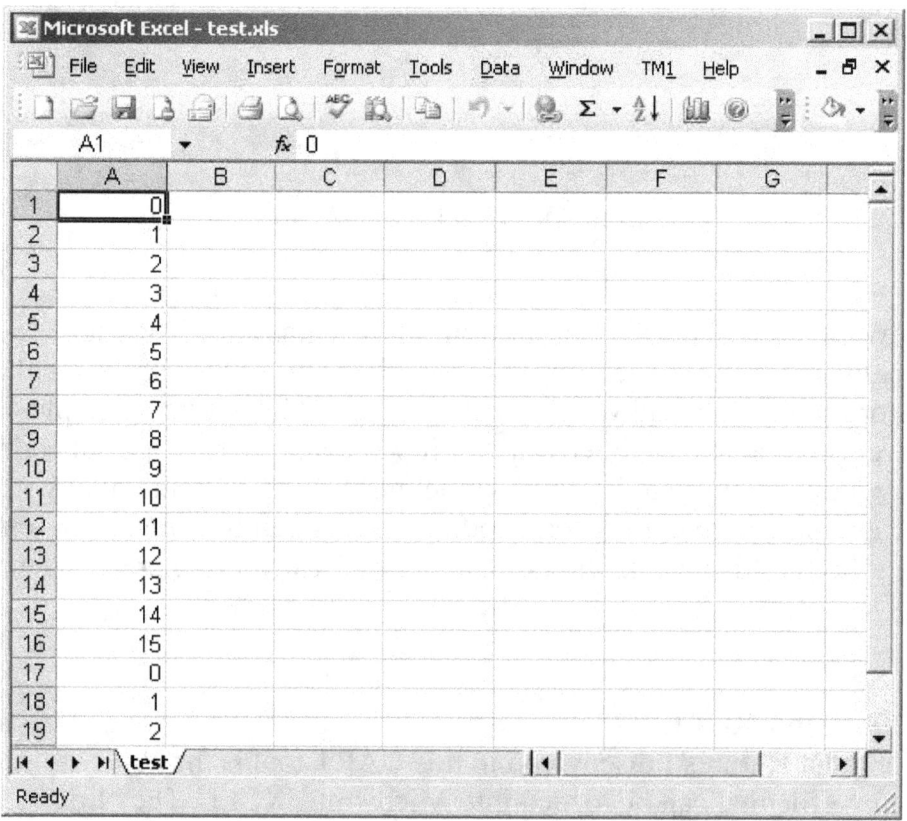

Figure 5 – EXCEL File with Data Logged Data

PICBASIC PRO Demo

I didn't mention that I compiled all this code with the new PICBASIC PRO demo code while running it in Microchip's MPLAB. This wasn't necessary since I'm programming from the standalone version of PICkit 2 software. You can write the code and compile it in any Integrated Design Interface (IDE) you want including the MCStudio version that comes with the PICBASIC PRO demo code when you download it. The reason I mention this is because the demo version also allows you to test running PICBASIC PRO in Microchip's MPLAB IDE and use the various features of MPLAB including the great simulator. That's an advantage because now you can write code for a smaller PIC MCU and test it on your PC without even hooking up hardware. Can it get any better?

By my count I only used 16 lines of code and I could have left a few of the setup commands off if I needed more space to fit within the 31 command limit. This whole setup worked so smoothly the first time I tried it I was driven to quickly write this up for you the reader. I also took an Atom chip and programmed it to send data serially out the same PICkit 2 pin but with the 28 pin demo board connected. It also worked great with the UART tool though I couldn't program the Atom with the PICkit 2 since the Atom programs itself via a bootloader. It just proved to me that the UART tool is a real handy addition to the PICkit 2 programmer. One of the reasons I used bootloaders in the past for all my development was to get this type of simple interface between the PIC MCU and the PC. I often would use a USB to RS232 adapter so I could use a laptop without a 9 pin serial port. The PICkit 2 delivers here again handling the USB interface.

This demo version should now give you beginners out there a great place to start small and see how much fun programming microcontrollers can be. A small investment in a PIC programmer and a few PIC MCUs plus the PICBASIC PRO demo should have you doing all kinds of simple but effective projects. I've gotten email from teachers who wanted recommendations for something beyond the $50 module path. Now this new setup should give the student/teacher no limit on how far they can take their programming skills.

Conclusion

I personally and publicly want to thank the great people at microEngineering Labs for the updated demo version. After you use it I know many will want to slide over to the full featured version so you can program any of the smaller pin count parts not to mention all the other PIC MCU's the PICBASIC PRO compiler supports. I often fail to mention that microEngineering Labs has their own USB programmer that works well with the MCStudio IDE. It also has numerous socket adapters for programming the various surface mount packages. I find their prices on these adapters are very reasonable. If you need surface mount you might look at the microEngineering Labs USB programmer as well.

I'm even looking into a programming contest using this new sample version of PICBASIC PRO so stay tuned for more of that. I'm also plugging away on my beginner book for programming in the embedded C language using this same programming tool path. That was the whole point of off-loading the hardware sales and development portion of elproducts.com to BeginnerElectronics.com. I just could not keep up with it all without something falling behind. My books were the one suffering and I have too many topics I'm working on to let it slip anymore.

You can visit my site www.elproducts.com for updates or details about my books and articles as I post them and please visit my friends at BeginnerElectronics.com to see how they are progressing. They've promised me they would continue to support the readers of this column along with my previous customers with the boards and designs I've used and continue to use in my books and articles. It's one of the things I hated when I started out. I would read about some new kit or project and then not be able to reproduce it because the kit or some parts were not available. BeginnerElectronics.com promises me that will not be an issue. Maybe I can convince them to offer a special reader discount. No promises but I'll keep you informed.

Also visit the microchip.com/pickit2 site for the latest PICkit 2 downloads and demo boards. This little programmer/debugger is really something special. Keep those emails coming because as this article proves, I listen to what you want and when I get a chance to talk to the big shots of this industry I will seize that opportunity. See you next month.

December 2007 - Stocking Your Lab

One of the more common e-mails I receive involves recommendations on which PIC® microcontroller (MCU) should be stocked in the home lab. Microchip offers so many choices, that the beginner can become very confused as to which ones to have handy. Engineers often also wonder if the choice of compiler matters, and then they get into questions about the best programmer, development boards, etc. to use. This month, I thought I would cover a topic I addressed before, but many have not read all my articles, so I thought I would cover my choices for the best microcontrollers to stock in your home lab.

Your choice of compiler won't matter much, unless you want to use one of the free sample or student versions. Even then, there might be only certain parts that the compiler supports. Microchip provides both 8- and 16-bit microcontrollers, as well as 16-bit Digital Signal Controllers (DSCs). In this column so far, we've focused on the 8-bit devices, so I'll continue that discussion here.

In Microchip's 8-bit family are the PIC10F, PIC12F, PIC16F and PIC18F prefix parts. The PIC10F and PIC12F are the small, 8-pin DIP package parts. The PIC10F gets even smaller, down to 6 pins if you will work with surface mount. As most hobbyists still develop with DIP packages, I recommend the PIC12F683 for the small 8-pin package size. The features of the PIC12F683 are listed below.

Figure 1: PIC16F683

PIC12F683 Features:
- 3.5 Kbytes/2 K-word program memory (14-bit address)
- 128 RAM
- 4 10-bit Analog-to-Digital Converters (ADCs)
- 1 Capture/Compare/Pulse-Width Modulation (PWM) peripheral

- 1 comparator
- 3 timers
- Internal oscillator up to 8 MHz
- 256-byte EEPROM

The PIC12F683 is also supported by microEngineering Labs' PICBASIC PRO™ compiler so, if you can write your code in 31 BASIC commands or less, this is a great part to get started with.

The PIC16F family is larger and includes package sizes from 14 up to 64 pins. For the 14-pin package, I recommend you stock the PIC16F688. This is Microchip's largest memory 14-pin package part and has all the features most of your projects will need. The features of the PIC16F688 are listed below.

Figure 2: PIC16F688

PIC16F688 Features:
- 7 Kbytes/4 K-word program memory (14-bit address)
- 256 RAM
- USART
- 8 10-bit ADCs
- 2 timers
- 2 comparators
- Internal oscillator up to 8 MHz
- 256 bytes of EEPROM

The next step up in the PIC16F family is the 20-pin package, and the best choice in my mind is the PIC16F690. The PIC16F690 has many of the same features as the PIC16F688, but more I/O. This is also the part that comes with the Microchip PICkit™ 2 Starter Kit that I've discussed in this article series before. Both the PIC16F688 and PIC16F690 are supported by the PICBASIC PRO compiler. The PIC16F690 is also supported by the HI-TECH PICC-lite C compiler that comes on the PICkit 2 Starter Kit CD. I liked this setup so much that I based my book on the

C programming language (book is still in the works) on this compiler and this PIC MCU. The PIC16F690 features are listed below.

Figure 3: PIC16F690

PIC16F690 Features:
- 7 Kbytes/4 K-word program memory (14-bit address)
- 256 RAM
- USART
- 12 10-bit ADCs
- 3 timers
- 2 comparators
- 1 Enhanced Capture/Compare/PWM (ECCP) peripheral
- Internal oscillator up to 8 MHz
- 256 bytes of EEPROM

I jumped over the 18-pin parts because the 8-, 14- and 20-pin parts share a common pin-out structure for the top 8 pins, so they can also share a development board. Figure 4 shows the structure of the common pins.

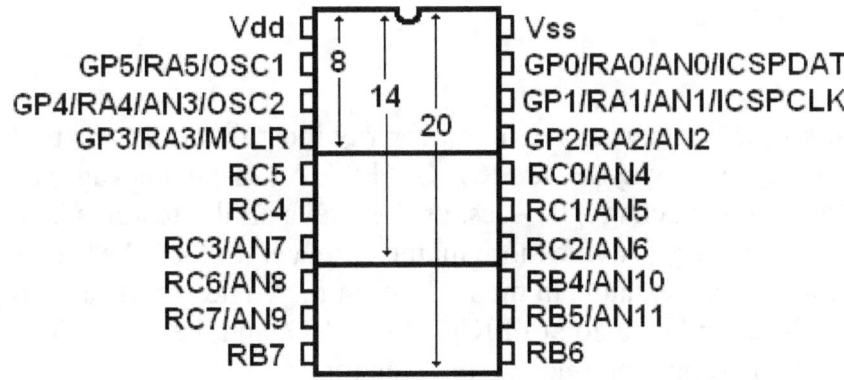

Figure 4: Common Pins

My recommendation for the 18-pin package is the PIC16F88, though I find I'm now using the PIC16F690 more often. The only thing you give up with this device is some RAM. If you have a program that requires quite a bit of variable space, then the 368 bytes of RAM in the PIC16F88 vs. the 256 bytes in the PIC16F690 can make a difference. The PIC16F88 also supports self-write memory, so you can program it via a serial port bootloader. This can be a nice feature.

Figure 5: PIC16F88

PIC16F88 Features:
- 7 Kbytes/4 K-word program memory (14-bit address)
- 368 RAM
- USART
- 7 10-bit ADCs
- 3 timers
- 1 Capture/Compare/PWM peripheral
- 2 comparators
- Internal oscillator up to 8 MHz
- 256 bytes of EEPROM
- Self-write Memory

The 28– and 40-pin packages have long been my most favorite parts to develop with. In the past, I've always used the PIC16F876A (28-pin package) and PIC16F877A (40-pin package) devices, as they have all the features and I/O I typically need, plus they provide more memory space. In fact, these have the largest memory space available in the PIC16F family. These parts are also supported by the sample version of PICBASIC PRO, and the PIC16F877A is supported by many other compiler sample versions.

I still recommend you keep some of these devices in your lab, but an upgrade to the family was recently released, and my two recommendations are the PIC16F886 (28-pin package) and the PIC16F887 (40-pin package). They add a few more features the PIC16F87XA parts don't offer. The biggest advantage is the internal oscillator. The PIC16F87xA parts require an external oscillator, but if all you need is a speed of 8 MHz or less, then the PIC16F88X parts have the same internal oscillator that the other parts I recommend have. So, I also recommend these parts for your lab.

Figure 6: PIC16F886

PIC16F886 Features:
- 14 Kbytes/8 K-word program memory (14-bit address)
- 368 RAM
- USART
- 11 10-bit ADCs
- 3 timers
- 1 Capture/Compare/PWM peripheral
- 1 ECCP peripheral
- 2 comparators
- 256 bytes of EEPROM
- Internal oscillator up to 8 MHz
- Self-Write Memory

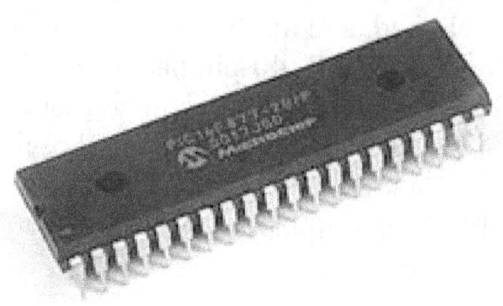

Figure 7: PIC16F887

PIC16F887 Features:
- 14 Kbytes/8 K-word program memory (14-bit address)
- 368 RAM
- USART
- 14 10-bit ADCs
- 3 timers
- 1 Capture/Compare/PWM peripheral
- 1 ECCP peripheral
- 2 comparators
- 256 bytes of EEPROM
- Internal oscillator up to 8 MHz
- Self-write Memory

PIC18F

Some people want a lot of memory to start with, so they don't have to worry about running out of room. The PIC18F family fills this request nicely. It includes a lot of parts and many with unique features such as motor control peripherals and built-in USB. However, for the hobbyist just looking for additional memory, I recommend you stock the PIC18F2620 and PIC16F886 devices. The 28-pin PIC18F2620 has 32 k words/64 kbytes of program memory. This is four times the space that the PIC16F886 offers.

Figure 8: PIC18F2620

The other features are larger as well, including EEPROM and RAM. The nice part is that the PIC18F2620 is pin compatible with the PIC16F886, so when your PIC16F design runs out of memory, then this PIC18F part can plug right in.

PIC18F2620 Features:
- 32 Kbytes/64 K-word program memory (16-bit address)
- 3968 bytes RAM
- USART
- Ten 10-bit ADCs
- 4 timers
- 2 Capture/Compare/PWM peripherals
- 2 comparators
- 1,024 bytes of EEPROM
- Internal oscillator up to 8 MHz

You can get a PIC18F4620 with the same features as the PIC18F2620 plus 3 additional ADCs and more I/O, but in a 40-pin package that is pin compatible with the PIC16F887.

Conclusion
There are many choices for the beginner looking to stock his or her home lab, but the parts discussed in this article will get you started for 99% of your projects. As I mentioned, Microchip also has 16-bit microcontrollers and DSCs in the form of the PIC24 MCUs, and dsPIC30 and dsPIC33F DSCs. These devices offer even more memory, with some very high-level features. The only BASIC compiler I know of for these higher-end parts is the mikroElektronica mikroB compiler, available at www.mikroe.com.

I use the PIC16F parts for 80% of my experiments, but I've been using the smaller-packaged PIC12F parts a little more lately. These devices give me so much capability in one package that easily fits into a small project box. Even if you just got a few of each of the parts discussed in this article you would have a great assortment in your home lab. You can order free samples at http://sample.microchip.com.

Please e-mail me at chuck@elproducts.com with any questions or comments. Please also visit my website www.elproducts.com anytime, and I'll offer as much help as I can.

Conclusion

The 12 months of articles collected for this book are the second year of the Nuts & Volts Magazine's "Getting Started with PICs" column and fortunately Nuts & Volts and especially the readers of this column enjoyed it enough to allow this column to continue for another year. I hope you'll find helpful tips and maybe a few key references in these articles that will make you look back at this book as a source of knowledge as you create your own electronic projects. If that occurs then this book was a success and my ideas were worth the hours of typing it took to create this book and the articles. Look for future books from me that collect future year articles as well. Offering the collected version allows those that missed an issue or those that just want their own printed copy can easily get what they are looking for. I have to admit I forget what I write and many times look at old copies of the magazine for reference so having this collection in a book form is helpful to me even if the book never sells.

You can always contact me at my email chuck@elproducts.com if you have comments or questions and also visit my website at www.elproducts.com to see what I'm up to or what book I may have released lately. All my hardware modules are now sold through my friends at BeginnerElectronics.com. I wanted to focus on my writing so I didn't have enough time to also design, assemble and sell the hardware. Programming microcontrollers and creating electronic gadgets is my passion. Helping others do it also is my goal and writing about it is my way of making that happen along with giving back the same way those that helped me did throughout my early years. Thanks for reading this book and see you in the pages of Nuts & Volts.

www.ingramcontent.com/pod-product-compliance
Lightning Source LLC
Chambersburg PA
CBHW081130170526
45165CB00008B/2626